AMORPHOUS SILICON

AMORPHOUS SILICON

Kazunobu Tanaka
National Institute for Advanced Interdisciplinary Research, Ibaraki, Japan
Eiichi Maruyama
Hitachi Ltd, Ibaraki, Japan
Toshikazu Shimada
Hitachi Ltd, Ibaraki, Japan
and
Hiroaki Okamoto
Osaka University, Osaka, Japan

Translated by
Takeshi Sato
National Institute for Advanced Interdisciplinary Research, Ibaraki, Japan

JOHN WILEY & SONS
Chichester • New York • Weinheim • Brisbane • Singapore • Toronto

Original Japanese edition published under the title *Ouyou Butsurigaku Shiriizu: Amorufasu Shirikon*, edited by the Japan Society of Applied Physics. Copyright © 1993 Kazunobu Tanaka, Eichi Maruyama, Toshikazu Shimada and Hiroaki Okamoto. Published by Ohmsha Ltd, Tokyo, Japan

English edition copyright © 1999 John Wiley & Sons Ltd,
Baffins Lane, Chichester,
West Sussex PO19 1UD, England

National 01243 779777
International (+44) 1243 779777
e-mail (for orders and customer service enquiries): cs-books@wiley.co.uk
Visit our Home Page on http://www.wiley.co.uk
or http://www.wiley.com

Other Wiley Editorial Offices

John Wiley & Sons, Inc., 605 Third Avenue,
New York, NY 10158-0012, USA

WILEY-VCH Verlag GmbH, Pappelallee 3,
D-69469 Weinheim, Germany

Jacaranda Wiley Ltd, 33 Park Road, Milton,
Queensland 4064, Australia

John Wiley & Sons (Asia) Pte Ltd, Clementi Loop #02-01,
Jin Xing Distripark, Singapore 129809

John Wiley & Sons (Canada) Ltd, 22 Worcester Road,
Rexdale, Ontario M9W 1L1, Canada

Library of Congress Cataloging-in-Publication Data

Amorufasu shirikon. English.
    Amorphous silicon / Kazunobu Tanaka ... [et al.] ; translated by
Takeshi Sato.
        p.   cm.
    Includes bibliographical references and index.
    ISBN 0-471-98293-8 (alk. paper)
    1. Silicon.   2. Amorphous semiconductors.   I. Tanaka, K.
(Kazunobu), 1940-   .   II. Title.
QC611.8.S5A4513   1998   98-7029
537.6′223–dc21   CIP

British Library Cataloguing in Publication Data

A catalogue record for this book is available from the British Library

ISBN 0 471 98293 8

Typeset in 10/12pt Times Roman by Laser Words, Madras, India
Printed and bound in Great Britain by Biddles Ltd, Guildford, Surrey
This book is printed on acid-free paper responsibly manufactured from sustainable forestry, in which at least two trees are planted for each one used for paper production

CONTENTS

3 Structural Properties   **58**

4 Optical and Electrical Properties   **103**

PREFACE TO THE
JAPANESE EDITION

'Et tu, Brute!'; so groaned Julius Caesar, when he recognized his beloved Brutus among a band of assassins in Shakespeare's *Julius Caesar*. I held the same feelings, though not so 'bitter', when in 1975 I first read a paper on the pn-control of amorphous silicon, and saw a breakthrough in it. This was the moment at which silicon took the leading role, not only in the world of crystalline semiconductors, but also in the realm of amorphous semiconductors.

Up to this time, studies on chalcogenide materials had constituted the main thrust of research on amorphous semiconductors, and amorphous silicon had little significance from the point of view of applications. At that time, I myself had been deeply attracted by the interesting properties, based on lone-pair electrons, of chalcogenide amorphous semiconductors and had plunged into the problems of photo-induced metastability of such materials.

At this stage, 'semiconductors' meant crystalline semiconductors. Silicon technology was a general term for semiconductor technologies represented by large-scale integration (LSI) based on crystalline silicon and had currently grown to a colossal entity, with LSI being regarded as the 'rice of industry'. Hence, those who had engaged in amorphous semiconductor research might well have been surprised with the predicted dominance of silicon in amorphous technologies. It should be noted, however, that amorphous silicon has substantially different properties as compared to crystalline silicon, and that research and development of amorphous silicon has taken place in an entirely different field, thus building up a technology of its own. This may be attributed to several factors.

First, a thin film of amorphous silicon contains a large amount of hydrogen which plays an essential role in the characteristics of such a material. In fact, this new technology is dealing with a 'new' material, different from conventional amorphous silicon. To be exact, this material should be designated as hydrogenated amorphous silicon (a-Si:H).

Secondly, the science of amorphous semiconductors which deals with structural and electronic properties is far from complete, being still in a developing stage, and in place of the conventional solid-state physics based on translational symmetry (long-range periodicity), the creation of a new form of physics for random systems is needed to complement the former. It could be said, therefore, that in the arena of amorphous semiconductors, 'you have no weapons for fighting, but you must fight while making the weapons'.

Thirdly, the plasma chemical vapor deposition (CVD) process for preparing a-Si:H involves problems of an entirely different nature from those of crystalline silicon. In particular, as a-Si:H is the first 'structure-sensitive' amorphous material, a perplexing diversity of structures and properties can result depending upon the preparative processes and conditions. For this latter reason, many of the pioneering groups fought desperately for reproducible results. The age has been asking for a preparative process based on microscopic analysis rather than conservative empiricism.

Against such a background, a-Si:H was thoroughly investigated from the materials point of view. Though not yet fully completed, a considerable amount of information has been accumulated in the past 17 years, covering the microscopic analysis of the preparative process, state of hydrogen inclusion, structures, defects and electronic properties. A new discipline of amorphous silicon physics has been created, injecting a breath of fresh air into materials science. In particular, studies on gas-phase processes centred on plasma diagnostics and surface properties have enabled researchers to examine the processes through which a material builds up its fundamental structures. This has provided a new concept and a new approach to studies on the materials synthesis which previously had fully depended on trial-and-error techniques.

On the other hand, multifarious developments occurred during this period in the realms of applications. Taking advantage of the relatively easy availability of large-area thin films, lower temperature processes at around 250 °C, and the continuous and extensive alteration of the photoelectric properties by alloying a-Si:H with germanium (Ge) or carbon (C), efforts have been continuously made to create innovative functions and reduce costs to a competitive level. Potential candidates for the application of a-Si:H include solar cells, thin-film transistor arrays for powering liquid crystal display panels, photoreceptors for copying machines, color sensors, etc. The rapid progress in research on a-Si:H owes much to the active and efficient exchange of results and experiences between those workers engaged in application and development and those concerned with the fundamental materials science. It is worth mentioning here that in Japan such an exchange has been made possible through the cooperation of the Japanese Society of Applied Physics.

Amorphous semiconductor physics has now reached adolescence. While it has grown smoothly up to now, it could be questioned as to whether or not it will become a fully fledged discipline of academic and industrial importance. For the sake of obtaining an insight into the future of this branch of amorphous semiconductor physics, it may be of particular significance to review amorphous silicon (a-Si:H) from its physics to its applications and then to put the outstanding problems in order from the academic point of view.

This book, *Amorphous Silicon*, has been produced as a new title in the 'Applied Physics Monograph Series' and has been co-authored by four specialists in the

field. The monograph focuses on hydrogenated amorphous silicon (a-Si:H). As
the physics of a-Si:H is still young and unrefined, attempts have been made to
compare the material with crystalline semiconductors in its structural aspects,
and to refer to its chalcogenide counterparts in the materials aspects, so that
its true nature may be more clearly visualized. In consideration of the fact that
the book has been written for specialists, the authors have tried to include not
only concepts and phenomena where their interpretation has been established
among the academic communities, but also those theories and techniques which
still remain to be resolved but seemingly hold significance from the scientific
and applications points of view, with the outstanding problems being explicitly
stated. As mentioned above, one of the characteristic features of a-Si:H physics
is the involvement of microscopic analysis in the preparative processes, thus
creating a new discipline of materials science. Taking this into consideration,
a separate section is devoted to 'Preparative Methods and Growth Processes'
(Chapter 2), and it is hoped that causal relationships may be made with the
'Structural Properties' which are discussed in Chapter 3.

Chapters 1–3, and 5 were written by K. Tanaka, Chapter 4 by H. Okamoto,
and Chapter 6 by T. Shimada and E. Maruyama; the manuscript was completed in
August 1992. Grateful thanks are due to Dr Akihisa Matsuda of the Electrotechnical Laboratory who provided valuable comments on Chapter 2. The descriptive
styles in the different chapters were coordinated as far as possible, but the authors
remain concerned about any remaining discrepancies, and inappropriate expressions which may be attributable to their misunderstanding or failure in citing
important works. In this respect, your criticisms and comments would be much
appreciated.

Kazunobu Tanaka
December 1993

Authors and Their Contributions

Chapter 1	Kazunobu Tanaka, Dr Eng.	National Institute for Advanced Interdisciplinary Research
Chapter 2	Kazunobu Tanaka	
Chapter 3	Kazunobu Tanaka	
Chapter 4	Hiroaki Okamoto, Dr Eng.	Osaka University
Chapter 5	Kazunobu Tanaka	
Chapter 6	Eiichi Maruyama, Dr Eng.	Hitachi Ltd, Project Leader, Joint Research Center for Atom Technology
	Toshikazu Shimada, Dr Sc.	Hitachi Ltd

PREFACE TO THE ENGLISH EDITION

This scientific monograph, 'Amorphous Silicon, has been translated from Japanese into English, to make it available to scientists and engineers throughout the world. This is a great pleasure for the author/editor of the original Japanese edition. The book was written jointly by four Japanese scientists, including myself, as one of the 'Specialists' Course texts of the Applied Physics Series published under the supervision of the Japanese Society of Applied Physics, with the first edition being published in March 1993.

This book describes the continuing development of hydrogenated amorphous silicon, ranging from the fundamental physics of amorphous solids, along with the relevant historical background, through the preparation, structural properties, optical and transport properties, up to various device applications. A particular emphasis has been placed on the materials science aspect of amorphous silicon, with a significant propotion of the text being devoted to the correlation between the film-growing processes and the properties of the resulting films. This latter aspect reflects the sizable contribution of Japanese scientists to this area. A large amount of research and development work, and studies of the applications of this material are in progress, so that the book may well deserve the alternative title, *Amorphous Silicon Technology*. Readers' comments and criticism on any aspect of the text would be most welcomed.

The translation from Japanese into English was entrusted to Mr Takeshi Sato of the Research Planning Office, National Institute for Advanced Interdisciplinary Research, and technical terms were checked and revised by the four co-authors for their own individual chapters.

Kazunobu Tanaka
Tsukuba, March 1998

1 INTRODUCTION

What is meant by the term 'amorphous'? From what background did amorphous silicon come in to the scientific world? How did it attract the attention of scientists? The aim of this introductory chapter is to provide the reader with intuitive replies to such questions. Comprehensive descriptions will be given for the definition of the amorphous state, the classification of amorphous semiconductors, the earlier history of amorphous silicon science, major breakthroughs and stages of development, and those researchers involved in this field, both domestic and overseas. Mention will also be given to the role played by various scientific associations in the development of amorphous silicon physics, as well as the establishment of a new academic community concerned with this field of study.

1.1 What is the Amorphous State?

1.1.1 DEFINITION OF TERMS

In the Introduction of the *Kojiki* (or 'Record of Ancient Matters', 712), which is one of the earliest documented histories of Japan, the author and court story-teller, Yasumaro Ohno begins his writing with the following:

> The chaos has begun to settle,
> But nothing has yet shown up,
> Neither naming nor display,
> Who can tell its morph?

This means that the nebulous primordiality of the universe has begun to condense, and something seems to taking shape. However, one cannot yet recognize any entity, nor describe its form [1]. This depicts the undefined primeval conditions before the gods appear, comparable to the Greek Myths which begin with the chaotic world. 'Amorphous' is the word to express such a state, i.e. 'chaos just beginning to settle'. The word is derived from the Greek language, i.e. α(= not) $+ \mu o \rho \phi \omega \varsigma$ (=shape), meaning 'without definite shape' or 'unclassifiable'. The amorphous semiconductor, which is dealt with in this book, is a 'solid without any definite shape' or a 'shapeless solid without any crystalline structure'. Solid-state physics has been created and built up since the beginning of this century as a science for dealing with crystalline materials in which atoms are arranged in an orderly manner. The term 'Solid' in solid-state physics has been used almost exclusively to refer to crystalline materials, while for non-crystalline materials, there have been neither any experimental means for analyzing their state, nor any theoretical approachs to describe their properties systematically, beginning with the first principle calculations. If solid-state physics based on crystals represents the world of myths ruled by gods, then amorphous solids represent chaos about to settle prior to the beginning of myths. Systematic studies of amorphous solids were initiated only in the 1970s, but at long last now seem to have acquired their citizenship in solid-state physics.

As will be discussed in detail later, an amorphous solid can assume a diversity of states, both thermodynamically and structurally, at certain fixed temperatures and pressures. This is significantly different from a crystalline solid which has fundamental structures which are uniquely defined at a given set of temperatures and pressures. Whether or not the structure can be determined experimentally, this characteristic may be regarded as the essential difference between crystalline and amorphous solids. Accordingly, if the state of the solid is to be described, it would be inevitable to classify solids into two categories, namely crystalline and its negative counterpart. In consideration of this fact, the following words may be used as synonyms for 'amorphous':

a-morphous	without definite shape
non-crystalline	without crystalline structure
dis-ordered	without ordered structure

It should be noted that the 'non-crystalline' state may sometimes include the liquid state. There is another set of synonyms with similar meanings:

random	orderless
glassy	glass-like
vitreous	glass-like

This second set has a restricted meaning for so-called glass-like substances (see Section 1.1.2). 'Vitreous' is derived from the Latin word '*vitrum*' (= glass). In Japan, modern glassware was brought by Portuguese and Dutch traders in around 1600 and called 'vidro' in Portuguese, which was also derived from *vitrum*. 'Glass', which has been used for designating glass ornaments or other objects since prehistoric time was derived from the word 'glesum' (= amber). Amber is a fossilized resin with a pale yellow color and transparent appearance. In ancient times, when microscopic analysis was not available, materials were 'character-ized' macroscopically by use of the human senses, particularly by vision. For this reason, glass and amber were regarded as being identical.

1.1.2 THE THERMODYNAMIC STATE

Let us now discuss the macroscopic state of amorphous solids in some detail. When ambient conditions such as temperature and pressure are given, materials take one of three states, namely solid, liquid and gaseous. In thermal equilibrium, 'solid' means the crystalline state, while both liquids and gases may be described as thermodynamically stable or metastable forms, with the exception of some special cases, such as ionized gases or electrolyte solutions. Certain special solids, as represented by the Al–Mn alloys discovered in 1984, exist as quasi-crystals, which may be defined as a metastable state [2, 3]. In the case of ternary alloys, a stable state can exist (see Table 1.1). If the state of a system is represented by its free energy G, then the latter can be related to the enthalpy H and temperature T by the following expression:

$$G = H - TS \qquad (1.1)$$

where H is related to the internal energy of the system and S to its state of disorder. The thermodynamically stable or metastable state corresponds to the presence of G at a least or minimum point, respectively. In contrast to a crystal (or quasi-crystal), liquid and gas, the thermodynamic state of an amorphous solid can be designated as a non-equilibrium state, i.e. the free energy G is not at its minimum. This situation will be described for glass, which is a typical amorphous solid.

In Fig. 1.1, the specific volume of a material is plotted as a function of temperature [4, 5]. If a liquid is cooled quasi-statically without disturbing the thermal equilibrium, it solidifies at a temperature T_m to form a crystalline structure (path a $\longrightarrow$ b $\longrightarrow$ c $\longrightarrow$ d in Fig. 1.1). This is a first-order phase transition which is accompanied by the release of heat (exothermic reaction). In contrast, if a solid is heated, it melts at T_m to turn into a liquid with heat being absorbed (endothermic reaction). T_m is the melting point, which is specific to the particular material. In other words, while the crystalline solid represents a stable state with the free energy of the system minimized at a temperature range

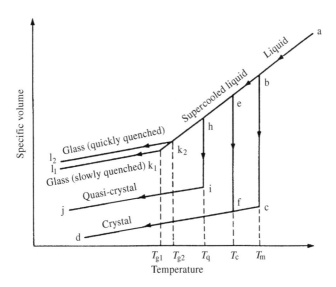

Fig. 1.1 Various states of a material represented in a specific volume versus temperature diagram

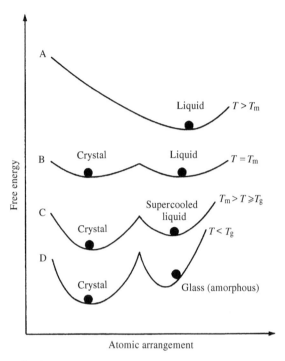

Fig. 1.2 Changes in the state of a material in response to variation in temperature as depicted in a free energy versus atomic arrangement diagram

below T_m, the stable state at temperatures above T_m is liquid. This means, that in equation (1.1), G of the system is minimized when reducing the enthalpy H at $T < T_m$, and also when increasing the entropy (or augmenting the disorder) at $T > T_m$. In Fig. 1.2, the free energy is plotted against the atomic arrangement for these various systems, while Fig. 1.3 shows two-dimensional representations of the corresponding atomic arrangements (A and C in both figures). When a liquid is cooled at an accelerated rate, it fails to crystallize at T_m as a result of its viscosity and becomes a supercooled liquid (path b $\longrightarrow$ k$_1$ in Fig, 1.1). On further cooling, the crystallization occurs at $T_c(< T_m)$ (path a $\longrightarrow$ b $\longrightarrow$ e $\longrightarrow$ f $\longrightarrow$ d in Fig. 1.1). The crystallization temperature, T_c, shifts to the lower temperature side, as the cooling rate is increased further. The crystallization of the supercooled liquid is represented by C in Fig. 1.2, by the movement from the right-hand trough (minimum value of G) to the left-hand trough (lowest value of G), thus overcoming an energy barrier.

If the cooling rate is increased still further, the liquid solidifies into another solid-state form without passing through the supercooled state (path a $\longrightarrow$ b $\longrightarrow$ k $\longrightarrow$ 1 in Fig. 1.1). This state of a solid, which is 'frozen liquid', as it were, which retains the disordered structure, is called a glass or an amorphous solid (B in Fig. 1.3). The temperature at which the molecular rotation and atomic diffusion

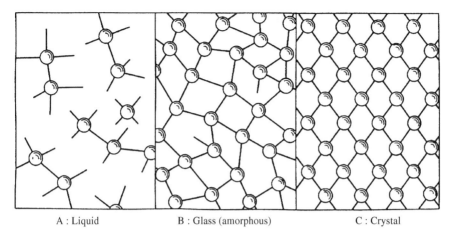

A : Liquid   B : Glass (amorphous)   C : Crystal

Fig. 1.3 Two-dimensional representations of various material states; circles denote atoms and lines chemical bonds

are frozen in a supercooled liquid is called the glass transition temperature, T_g. As T_g shifts only slightly (depending upon the cooling rate), this phase change is generally not considered to be a secondary phase transition. If the cooling rate is extremely high, the glass transition temperature shifts to the higher-temperature side, i.e. $T_{g1} \longrightarrow T_{g2}$. It is logical, therefore, to regard amorphous solids, which change in surprisingly diversified manners depending upon the cooling rates, as 'non-equilibrium states', thermodynamically. The free energy G of the glass (amorphous) system is characteristically displaced from the minimum point when frozen at T_g, as shown by plot D in Fig. 1.2.

To summarize, an amorphous solid is a non-equilibrium state of a material which can take a variety of different macroscopic states depending upon the preparative process (see Table 1.1). In view of the extremely long time-scales involved, far exceeding those occurring under normal laboratory conditions, the macroscopic state must be able to change very slowly. The entropy is held fixed at a level as high as that in liquids, with the enthalpy being fixed at a much higher level than that found in crystals. As a reference, the behavior of quasi-crystals, such as the Al–Mn alloy systems, is illustrated in Fig. 1.1 (path a $\longrightarrow$ h $\longrightarrow$ i $\longrightarrow$ j).

1.1.3 STRUCTURAL DISORDER

As discussed above, amorphous solids imply materials at a higher level of disorder. Therefore, how should we describe the degree of structural disorder or irregularity? Generally speaking, it is by no means easy to define disorder. Let us try to set a standard for the orderly arrangement of atoms and then identify some hierarchical levels of deviation from this standard. In other words, the degree of disorder is to be measured by reference to the crystalline or quasi-crystalline arrangement.

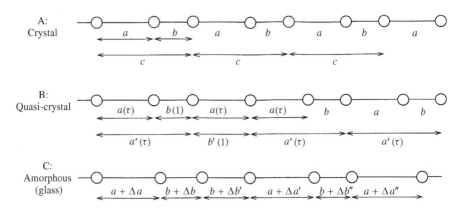

Fig. 1.4 One-dimensional representations of the atom arrangements in three different material states

In Fig. 1.4, conceptual illustrations of various one-dimensional models are presented for the crystalline, quasi-crystalline and amorphous states, in which circles denote the atoms, and a and b represent the interatomic distances. First, let us assess the degree of orderliness in terms only of the relationship between the adjacent atoms. In the crystalline and quasi-crystalline states, both a and b are always fixed. This may be described in the following way: the short-range order (SRO) is fully maintained. On the other hand, in the amorphous state both a and b are not constant, with some local fluctuations Δa and Δb being present. However, the relative fluctuations $|\Delta a|/a$ and $|\Delta b|/b$, are not that large, because element-specific chemical properties persist in the relationship between nearest-neighbour atoms even in the amorphous solid. Therefore, it may be claimed that a type of near-crystalline SRO is present, although slightly imperfect in nature.

However, if we now consider two atoms which are located, say a dozen atomic distances apart, substantial differences reveal themselves in the crystalline, quasi-crystalline and amorphous states.

In the case of the true crystal, where a and b alternate constantly and endlessly, an infinitely extensive crystal can be constructed by repeated parallel shifts of adjacent pair of atoms by an integral multiple of $a + b(= c)$. This kind of order is termed as translational symmetry. The system is of a periodic structure, and is composed of diatomic unit cells with a lattice constant c. Hence, there is perfect order, even between atoms which are distantly separated from each other. This is generally called long-range order (LRO).

This long-range order exists in quasi-crystals, too, although the two unit cells with dimensions a (longer) and b (shorter) are by no means arranged in regular periods. In Fig. 1.4(B), the position of the nth atom is given by the following expression [3]:

$$\chi_n = n + \frac{1}{\tau}\left\lceil\frac{n}{\tau}\right\rceil \tag{1.2}$$

where [ ] denotes the Gaussian symbol, $[\chi]$ stands for the largest integer below χ, and τ is the reciprocal of the 'golden section' $(= (1 + \sqrt{5})/2)$.

In Fig. 1.4(B), $a = \tau$ and $b = 1$, and the sequence of spacings between the adjacent atoms is not periodic, such as $\tau 1 \tau \tau 1 \tau 1 \tau \tau 1 \tau \tau 1 \tau \ldots$, etc. This is called a Fibonacci sequence, and is self-similar with respect to the transformations $1 \longrightarrow \tau (a \longrightarrow b')$ and $\tau \longrightarrow \tau 1 (a + b \longrightarrow a')$ (see Fig. 1.4(B)). In other words, while both crystals and quasi-crystals have long-range order (LRO), the former is characterized by periodicity, while the latter displays a non-periodic self-similarity.

In contrast, amorphous solids have no long-range order. Any persistent order is eliminated when atoms dozens of atomic distances apart are compared, as the small fluctuations Δa and Δb are integrated into the system.

To summarize, the structural randomness of amorphous solids is characterized by the lack of long-range order, while short-range order is adequately maintained. This condition closely resembles that of a liquid. This is another way of saying 'freezing a liquid leads to glass'.

Let us now consider the short-range order of amorphous solids more specifically by using a three-dimensional network model involving silicon or germanium. Since our amorphous solid is based on a covalent-bond (or electron-pair-bond) structure, each of the silicon or germanium atoms, for example, is equipped with four covalent bonds based on sp^3 hybridized orbitals for building up a three-dimensional network (see Fig. 1.5). As factors for determining the short-range order, fluctuations in the covalent bond length (r_1), the bond angle (θ), and the second neighbor distance (r_2), are usually taken into consideration, while the third neighbor distance (r_3), and the dihedral angle (ϕ) are discussed separately for intermediate-range order (IRO) [6] (see Fig. 3.11 in Chapter 3). It should be noted here that in the physics of amorphous solids,

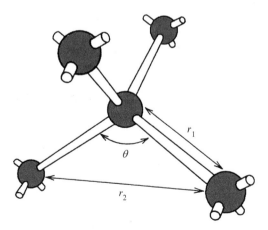

Fig. 1.5 Illustration of short-range structural order based on sp^3 hybridized orbitals, showing the parameters that need to be considered

Table 1.1   The various states of materials and their corresponding features

Material state	Thermodynamic state	Atomic arrangement[a]		
		Short-range order	Long-range order	Translational symmetry
Crystal	Stable	◎	◎	◎
Quasi-crystal[b]	Metastable[e]	◎	◎[f]	×
Amorphous	Non-equilibrium	○	×	×
Liquid[c] (supercooled liquid)	Stable (metastable)	○	×	×
Gas[d]	Stable	×	×	×

[a] ◎ = almost perfect order; ○ = imperfect order; × = disorder
[b] No bulk quasi-crystals have yet been found in semiconductors
[c] With the exception of electrolyte solutions
[d] With the exception of ionized gases
[e] Stable ones have recently been found
[f] Long-range order based on self-similarity

the order intermediate between the short-range order and the long-range, i.e. the intermediate-range order, plays a significant role. This type of order is discussed in detail in Chapter 3.

The discussion so far is summarized in Table 1.1. In addition to the above characteristics, alloy systems may involve the problem of entropy of mixing, while thin films may be affected by heterogeneity over much larger areas than discussed here and also by microvoids. These aspects will be discussed more fully in Chapter 3.

1.1.4 CLASSIFICATION OF AMORPHOUS SEMICONDUCTORS

In the preceding section, the structural and thermodynamic features of amorphous solids have been described in detail. As with crystalline solids, amorphous solids can be classified into various categories depending upon their properties. Amorphous semiconductors may be defined as 'amorphous solids possessing semiconductor features', and then further subdivided into subcategories in accordance with their structures and functions.

If a semiconductor is defined, for the moment, in an extended sense, as 'a material capable of electronic conduction with the conductivity of $10^3 - 10^{-10}$ Ω^{-1} cm^{-1} at room temperature where the latter increases with the temperature T, in contrast to the conductivity of a metal', amorphous semiconductors may be classified as shown in Table 1.2. This classification may be based on various features such as the groups in the periodic table to which the main component elements belong, the type and structure of the chemical bonds, the coordination number of the nearest neighbors, preparative process, etc. [7]. However, as these features are closely related to one another, the most well-known, generalized designation has been used.

The designation and classification used in Table 1.2 is based on groups in the periodic table. In the tetrahedrally bonded category, atoms of group IV

Table 1.2 Classification of amorphous semiconductors

Designation	Examples of amorphous materials	Main component	Coordination number of main component	Preparative process
Tetrahedrally bonded	Si, Si(H), Ge(H), Si_xC_y, Si_xGe_y, Si_xSn_y, GaAs	Group IV elements (Si, Ge) Group III–V elements	4	Gas freezing
Pnictide	P, As	Group V elements	3	Gas freezing, liquid freezing
Chalcogenide	Se, As_2Se_3, As_2S_3, As_2Te_3, As–Te–Ge, As–Se–Ge–Te	Group VI elements (chalcogens), S, Se, Te	2	Gas freezing, liquid freezing
Oxide	V_2O_5–P_2O_5–BaO, MoO_3–P_2O_5–BaO	Group VI elements (oxygen), transition metal	2	Liquid freezing

elements with four electrons in the outermost shell ($N = 4$) have sp^3 hybridized orbitals and build up a rigid three-dimensional network based on the tetrahedral coordination of strong covalent bonds. The pnictide and chalcogenide semiconductors have group V and VI elements, respectively, as their main components. According to the $(8 - N)$ rule, or octet theory, the number of bonds per atom (i.e. the coordination number) is 3 in the pnictide, and 2 in the chalcogenide materials. If the coordination number is smaller, the network structure tends generally to be reduced in dimensions to layer-like and chain-like configurations. However, the intermediate-range order varies in a complicated manner depending upon the preparative process. The relationship between the coordination number and the preparative process will be described in Chapter 2. A chalcogen element has six electrons in its outermost shell, i.e. s^2p^4, of which the two s-electrons are localized in a deep-energy state in the atom, and the two p-electrons form two covalent bonds, thus giving a coordination number of 2. The remaining two p-electrons form no bonds, but retain their lone pairs. The energy of lone-pair electrons is higher than that of covalent electrons, and provide the top level of the valence electron band. The lone-pair electrons interact with adjacent atoms in a complicated manner to produce the characteristic properties of chalcogenide materials.

The oxide semiconductors contain a group VI element (oxygen) as the main component. The energy gap of the oxide is much higher than that of the chalcogenide, and its conductivity based on ionic conduction is as low as that of insulator. Table 1.2 lists special oxide-based glasses with high conductivities (including electronic conduction) containing transition metals. These oxide semiconductors are distinguished from chalcogenide materials because they are qualitatively different with respect to both their preparative processes and their conduction mechanisms.

1.2 A Short History of Amorphous Semiconductor Studies

1.2.1 PIONEERING WORK

The chalcogenides were the first amorphous semiconductors to attract attention as functional materials. As with the history of crystalline semiconductors, which started with crystalline selenium (Se) and cuprous oxide used for rectifiers, the origin of amorphous semiconductors can be traced back to amorphous Se involved in xerography. The selenium described in the paper on xerography by Schaffert and Oughton [8] is reported to have a dark resistivity of 10^{15} Ω cm and a light resistivity of $10^{10}-10^{12}$ Ω cm. It is doubtless that this selenium was amorphous Se [8]. For these workers, the xerographic function was the matter of significance regardless of the microscopic structure of the material, and no mention was made of 'amorphous' in their paper. As far as this author is aware, this was the first example of an amorphous semiconductor being used as a functional material. This same year, i.e. 1948, is also famed for the announcement of the point-contact germanium transistor.

The first paper to appear which made reference to the amorphous nature of the material 'Photo-Conductivity in Amorphous Selenium' was published in 1950 by Weimer of RCA [9]. Weimer pointed out a wrong conclusion that had been drawn in the previous literature that 'no photoconductivity exists in amorphous materials' through his own measurements of the spectral sensitivity of photoconduction. Subsequently, a considerable amount of work has been accumulated on amorphous selenium concerning both xerography and photoconductors. Spear, who was to later successfully realize pn control of amorphous silicon (Si) for the first time, initiated research on amorphous selenium at around 1955 in order to follow the carrier generating process [10]. However, he was not aware of 'semiconductors' and all materials were therefore designated as 'insulators'.

The prototype amorphous semiconductor was created at the Ioffe Physical Technical Institute in Leningrad (now St. Petersburg), Russia. B.T. Kolomiets, born in 1908, started his studies of photodiodes in 1930, under A.F. Ioffe, 'the father of Russian semiconductors'. While in the earlier half of the 1930s, photodiodes based on cuprous oxide or crystalline Se had photoelectric conversion efficiencies as low as 0.01 to 0.04%, Ioffe's laboratory in 1937 succeeded in dramatically improving the efficiency to 1.1% with the binary crystalline semiconductor Tl_2S. Subsequently, the advent of Si and Ge in 1952 boosted the efficiency to 14%, whereas Kolomiets persisted with chalcogenides in an attempt to extend the range of materials to include ternary systems. He began with 1:1 compositions of binary crystals such as Tl_2S, Tl_2Se, Sb_2S_3 and Sb_2Se_3. It was found that when Sb was substituted with As, and a mixture of Tl_2Se and As_2Se_3 crystals was melted to prepare ternary systems, a glass-like structure was obtained, with the conductivity σ turning out to be as high as 10^{-6} Ω^{-1} cm^{-1}, which was several orders of magnitude higher than that of amorphous Se. This was the first amorphous semiconductor that Kolomiets worked with and in that same year (1955) he was appointed to be head of the Laboratory of Photoelectric

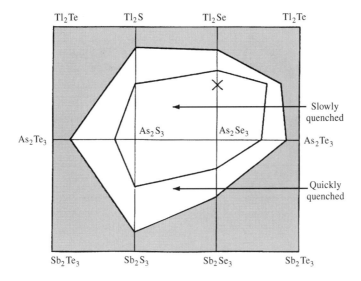

Fig. 1.6 The composition of the various chalcogenide glasses originally prepared by Kolomiets and coworkers (×) and the area of vitrification; the shaded area represents the region of non-vitrification (crystallization) [12, reproduced with permission]

Phenomena of the Ioffe Physical Technical Institute [11]. By extensively altering the composition, a wide area of vitrification was found (as shown in Fig. 1.6). Kolomiets and coworkers named this group of materials as chalcogenide glasses, and described the following properties on the basis of a series of systematic experiments.

(1) In contrast to many of the other glass systems which display ionic conduction, chalcogenide glasses display electronic conduction.

(2) The electrical conductivity σ of chalcogenide glasses can be as high as 10^{-3} Ω^{-1} cm^{-1}; its temperature dependence, which is of the thermal activation type, as found for intrinsic crystalline semiconductor, is given by the following expression:

$$\sigma = \sigma_0 \cdot \exp\left(-\frac{\Delta E}{kT}\right)$$

(3) There is an optical absorption edge.

(4) The thermoelectric power is high.

(5) The photoconductivity (internal photoelectric effect) is of a significant magnitude.

Since these properties are also observed in crystalline semiconductors, these materials were named for the first time, vitreous semiconductors, in order to distinguish them from crystalline semiconductors, [12].

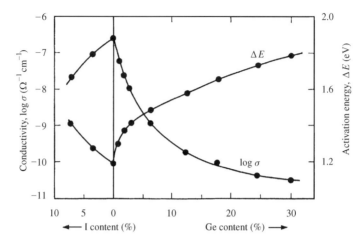

Fig. 1.7 Results of the first doping experiments of $Tl_2Se-As_2Se_3$ glass with impurities (iodine and germanium) [12, reproduced with permission]

The most important experiments carried out by Kolomiets' group, however, appears to be doping (with impurities), as shown in Fig. 1.7. By analogy with crystalline semiconductors such as Si and Ge, they added minute amounts of impurities such as Ge (group IV) or I (group VII), to Tl_2Se and As_2Se_3 (groups III, V and VI), respectively, and measured the changes in their electrical properties.

As shown in Fig. 1.7, when Ge (germanium) or I (iodine) is added at concentrations up to the order of *ca* 1%, the conductivity σ changes by only an order of magnitude, in contrast to crystalline semiconductors where σ changes by a few orders of magnitude with additions of dopant of the order of ppm. This experiment is of historical significance, demonstrating that vitreous semiconductors at this stage were not structure-sensitive. For two decades, many workers challenged the significance of doping vitreous semiconductors, until the success later achieved by Spear and coworkers in the 1970s.

This author (KT) has happy experiences of talking informally with Dr Kolomiets on three occasions, namely at the Leningrad International Conference in 1975, at the Ioffe Institute in 1981, and at the Prague International Conference in 1987. To a question 1 presented on his motivation for starting his studies on chalcogenide glasses following his previous work on ternary crystals, he replied 'I wanted to step into an area where no one had yet poked their nose into.' 'At this time, when crystalline semiconductors, such as Si and Ge, were about to enter their heyday, a person who deals with materials with very low electron mobilities could be regarded as being eccentric. His tutor, Dr Ioffe was the only person who supported and encouraged his intentions. Dr Kolomiets, who devoted himself to the study of chalcogenides, struggling against an adverse tide, as it were, had both defying looks and personality, and was as energetic and enthusiastic as to regularly scold loudly his followers, even at the age of 73 in

1981. It may be interesting to point out, in portraying his personality, that he continued to be 'cold' to hydrogenated amorphous silicon, which first stepped on to the stage in 1975. Regrettably, he died in October 1989.

The systematic studies by the Kolomiets group were not so impressive as to attract that many researchers' interest in the face of the rising interest in crystalline semiconductors, although they did trigger some important problems. For example, while the band theory based on the Bloch conditions is not applicable to solids without long-range order, why is there an optical absorption edge in these glasses, which suggests the band gap?

The first theoretical guidelines for amorphous semiconductors were given by Ioffe and Regel in 1960 [13]. In an article entitled 'Non-crystalline, amorphous and liquid electronic semiconductors' they proposed the following:

> The fundamental electronic property of a solid, i.e. whether it is a metal, semiconductor or insulator, is determined principally by the short-range order of its constituent atoms, or by the coordination number of the nearest neighbour atoms.

This general empirical rule was derived from the theory of liquid metals which had been built up over the period from the 1930s to the 1950s, of which the leading figures included N.F. Mott, as well as from an enormous amount of experimental data. The rule points out that there can exist an equivalent of the band gap in terms of chemical bond theory, even if long-range order is lacking. This concept was theoretically verified by Weaire and Thorpe in 1971 [14], where they demonstrated exactly, by using a very simple model Hamiltonian, that the energy gap can evolve from the short-range order alone.

N.F. Mott, who received the Nobel Prize for physics in 1977 for his work on 'The Electron Theory of Disordered Systems', had already started his work on disordered systems in the 1930s with studies of the theory of liquid metals. According to his commemorative speech given when receiving the Nobel Prize [15], he began to take an interest in the disordered solids in the 1960s, which had been triggered by the question as to why glass was transparent, and by a series of experiments carried out by Kolomiets and his group. Mott's theoretical work will be mentioned in various parts of this book, and are not discussed any further here. Instead, a few passages are presented from his Nobel Prize speech in order to illustrate the significance of his studies:

> What I did in this field was to look into every experimental fact, to calculate something on the back of an envelope or something like that, to give theorists a prophecy that by applying my methodology to this problem would yield such and such results, and to instruct experimentalists in a more or less similar way.[†]

In any way, Professor Mott is a living witness to the half-century evolution of disordered system problems, and remains an octogenarian (he was born in 1905) patron of amorphous semiconductor physics.

[†] This is not a direct quotation from his speech, but a 'back' translation from the Japanese translation by Ms Fumiko Yonezawa

P.W. Anderson, a corecipient, with Mott, of the Nobel Prize for physics in 1977, is famous for his work on diffusion in random lattices, which is the starting point for the so-called 'Anderson localization' [16] (see Section 4.1.3). Anderson made another important contribution to the world of amorphous semiconductors through his proposal of the epoch-making model for localized levels [17], which will be discussed in detail in Chapter 5 (see Section 5.2.2).

Following the guidelines proposed by Ioffe and Regel [13], amorphous Se was now being considered as 'an amorphous semiconductor'. A group of scientists at the University of Illinois, headed by J. Bardeen, who has twice received the Nobel Prize in physics, namely for the invention of the transistor and for the theory of superconductivity i.e. the Bardeen-Cooper-Schrieffer (BCS) theory, in 1962 presented an energy-distribution model for the bandgap and localized levels in amorphous Se, based on experiments on drift mobility and space charge-limited current [18] (see Fig. 1.8). This was probably the first band diagram to appear in the literature. As was pointed out by Bardeen in this paper, discrete high-density energy levels, both above and below a bandgap, may be replaced by the continued and gradual transition of the energy ends of the valence and conduction bands without any sharp edges. This concept is inherited in the mobility-gap model for amorphous semiconductors which was proposed by Cohen, Fritzsche and Ovshinsky in 1969, the so-called CFO model [19], thus giving the energy gap in amorphous materials an increasingly definite shape. This model will be discussed in detail in Chapter 4.

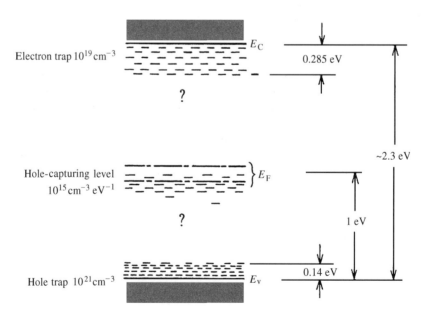

Fig. 1.8 A schematic diagram for the band model of amorphous Se [18, reproduced with permission. Copyright 1962 by the American Physical Society]

In 1968, a paper on the electrical switch/memory effects in thin films of chalcogenide amorphous semiconductors (the As−Te−Ge−Si system) appeared in the November 11 issue of *Physical Review Letters* [20], and caused a great sensation. In spite of the obscurity of the author, S.R. Ovshinsky, in the physics community a press announcement was also issued with favourable comments added by such distinguished scientists as N.F. Mott and H. Fritzsche to advocate the possibility of replacing the crystalline transistor with an amorphous one. The switch/memory effect itself had been reported in a more or less similar form in 1962 [21], but failed to exert any significant impact in the applications field. However, nobody could deny the fact that this paper provided an opportunity for many scientists to pay attention to amorphous semiconductors and gave a primary impact for the augmentation of research work in this field. It seems that Ovshinsky asked Bardeen for his comments before publishing his work on the memory/switch effect, and then met Fritzsche on the introduction by Bardeen [22]. Fritzsche stepped into the world of amorphous semiconductors at this moment and since then the physics of amorphous semiconductors has continued to make rapid and steady progress.

Now, let us review the studies on amorphous semiconductors in Japan. The Japanese Society of Ceramics initiated research on chalcogenide glasses and oxide glasses from the latter half of the 1950s into the 1960s. The major research topics involved the optical properties of chalcogenide glasses [23], and the mechanism of electron conduction of oxide glasses containing transition metals, with respect to non-ionic conduction [24]. However, no attempt had been made to conduct systematic studies from the view point of functional materials (semiconductors) or the physics of disordered systems. The population of Japanese researchers in this field had continued to grow since 1968 when Ovshinsky's paper appeared. In the earlier half of the 1970s, in particular, Japanese physicists had built up their own specific research fields, covering the photostructural changes in chalcogenide systems [25−27], the photo-doping of silver [28], and the latter's application to photoresists [29].

I would like to refer here to two contrasting works carried out in Japan before the publication of Ovshinsky's paper. The first of these concerns the collaborative development of the television image pick-up tube by Hitachi Ltd and NHK (the Japan Broadcasting Corporation). The project started in 1965 with an image pick-up tube using amorphous a-As_2Se_3, and after a decade of continuing trials, culminated in the commercialization of an image pick-up tube based on the Se−Te−As system [30]. Although the potential application of a-As_2Se_3 had been already suggested by both the Kolomiets group [12] and the RCA group [29], the Japanese groups took a serious risk by throwing considerable human resources and time into this project before an independent community of amorphous semiconductor physics had been established. In contrast to the RCA group having announced what were somewhat pessimistic prospects, the patient efforts for success of the Japanese scientists should be clearly appreciated [30]. The other work is an example of studies based on a creative idea, but which had failed to

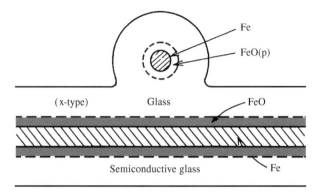

Fig. 1.9 A cross-section of a glass junction diode constructed by using semiconductive glass containing a transition metal (e.g. Fe) [31, reproduced with permission]

make an impression in people's minds at the time. In 1965, Yamamoto of the Institute of Physical and Chemical Research had produced a device (structured as shown in Fig. 1.9) using oxide glass containing a transition metal (vanadium) by taking a suggestion from the work of Munakata and coworkers of the Electrotechnical Laboratory [24] and demonstrated negative resistance [31], as shown in Fig. 1.10. Moreover, as is evident from Fig. 1.9, if a number of iron wires are laid in parallel and in dual layers in a criss-cross fashion, an array of switching elements is produced. Yamamoto called this arrangement a 'glass fibre plan' in his paper published in the *Journal of the Physical Society of Japan* [31]. This was regarded as a unique and original idea by the community of physicists in Japan at this time. His work failed, however, to attract other people's attention and faded away without any significant follow-up. If this work had been adequately

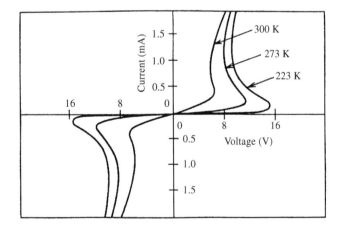

Fig. 1.10 Negative-resistance characteristics of the glass junction diode shown in Fig. 1.9 (with iron replaced by vanadium) [31, reproduced with permission]

followed up after the developments of 1968, some interesting results might have been obtained, although this is a somewhat hypothetical suggestion.

1.2.2 ADVENT OF HYDROGENATED AMORPHOUS SILICON

While studies of chalcogenides constituted the main thrust of research in amorphous semiconductor physics in its pioneering period, a few outstanding works were reported on tetrahedrally bonded materials. R. Grigorovici (Rumania) in 1964 observed a rectifying effect in heterojunctions of crystalline Ge with amorphous Ge (a-Ge), thus indicating that a-Ge is not structure-sensitive [32]. This was the first report on tetrahedrally bonded materials from the view point of a semiconductor. In 1965, Tauc (Czechoslovakia, who later emigrated to the USA), a coworker of Grigorovici, announced some specific optical properties of a-Ge, which later became known as 'Tauc's plot' [33]. Subsequently, Grigorovici examined an ideal structural model of a-Ge by using the Voronoi polyhedron, and proposed the structural unit, known as an 'amorphon', shown in Fig. 1.11, thus suggesting the importance of fivefold symmetry [34]. An amorphon, which is a regular dodecahedron, is closely related to the fivefold symmetry of the icosahedron, which was one of the essential concepts of quasi-crystals to come to significance decades of years later [2]. However, discussions of the structure were somewhat confused in the absence of sufficient experimental data, and no satisfactory correspondence was established between experiment and theory until 1970 (this will be discussed in detail in Chapter 3). It should be noted, however, that the study of amorphous semiconductors originated in the former USSR and other Eastern European countries for both the chalcogenide and tetrahedrally bonded materials.

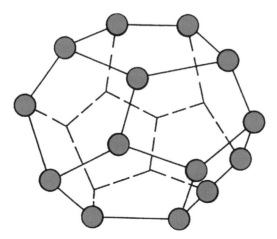

Fig. 1.11   The amorphon structure proposed as an ideal structure for a-Ge [34, reproduced with permission]

Following Ovshinsky's paper, the study of a-Ge and a-Si became internationally popular from the latter half of the 1960s through to the earlier half of the 1970s. During this period, samples were prepared mainly by sputtering and evaporation, which both required high vacuum facilities for making purified materials. In the midst of such a climate, two groups stuck stubbornly to the preparation of a-Ge and a-Si by using low-level evacuation systems, (oil rotary vacuum pumps), namely R.C. Chittik and coworkers, and a group led by W.E. Spear and P.G. LeComber. These workers relied on the glow discharge technique, where GeH_4 or SiH_4 is decomposed by a radio-frequency (RF) glow discharge (the plasma chemical vapour deposition (CVD) method, see Chapter 2).

In 1969. R.C. Chittik and coworkers prepared amorphous Si (a-Si) by the RF glow discharge technique, and measured its electrical and optical properties, using the growth conditions (i.e. substrate temperature) as a parameter [35]. When a member of this research group, H.F. Sterling had already attempted 1965 to grow a-Si via glow discharge [36], Chittik was the first person to try systematic characterization[†]. The a-Si specimen prepared by glow discharge showed certain distinct characteristics in comparison to a-Si grown by other methods. As shown in Fig. 1.12, the former had a very high photosensitivity. Furthermore, a temperature dependence of the resistivity was of a distinct activation type, with activation energies as high as 0.5–0.8 eV. Chittik's group, who conducted their doping experiments by using SiH_4 mixed with PH_3, missed a precious chance by a narrow margin, because of their inadequate checkmate strategy [35]. A challenge to the 'structure-sensitive' behavior thus seemed to fail once again, but the experiments by Chittik's group were destined to prime the successes later achieved by Spear and others, as will be described below.

Various theoretical explanations for the 'structure-insensitive' nature of these materials were published independently in the 1960s by Gubanov [37] and Mott [38]. The former pointed out that the electrons and holes supplied by the impurities were captured by a number of localized levels in the bandgaps derived from the structural randomness, thus suppressing any extensive shift of the Fermi level. The latter claimed that impurities with different valences managed to meet their valency requirements owing to higher structural degrees of freedom, thus cancelling the feeding of electrons and holes. It could be supposed that in the actual materials these two mechanisms work together. For further details, see Chapters 3 and 4. Although these theoretical approaches do represent some attempts to give reasonable explanations for the existing experimental data, sometimes it happens that these are mistaken as being absolute laws. This is an unfortunate pitfall which is often encountered, particularly when the theory is advocated by an influential scholar such as Mott. The breakthrough

---

[†] During World War 2, a project on anticorrosion treatments for spark plugs in military vehicle and aircraft engines was carried out in 1942 at the Noguchi Laboratory in Yokohama by coating the plugs with amorphous silicon through the decomposition of SiH_4 or Si_2H_6 via glow discharge. This information was obtained by the author (KT) directly from Mr Fujio Shiga, one of the participants in the project.

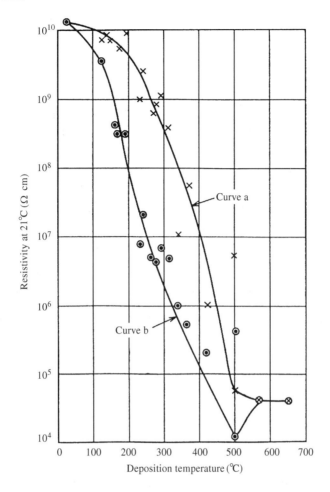

Fig. 1.12   The deposition temperature dependence of the resistivity of amorphous silicon prepared by the glow discharge method, in the dark (curve a) and in light (curve b). [35, reproduced with permission]

experiments which escaped such a blind alley was carried out by Spear's group at the University of Dundee, Scotland in 1975, two decades after the pioneering experiments by Kolomiets.

Spear and coworkers used as specimens thin films of a-Si prepared in the same way as Chittik's group, i.e. by the glow discharge technique (by decomposing SiH_4 in a plasma). Stimulated by the work of Chittik and coworkers, they had been systematically accumulating data using such specimens since around 1969, disregarding amorphous silicon prepared by evaporation and sputtering. Their data, which are of considerable historical significance [39], are shown in Fig. 1.13. This figure clearly demonstrates the structural sensitivity caused by

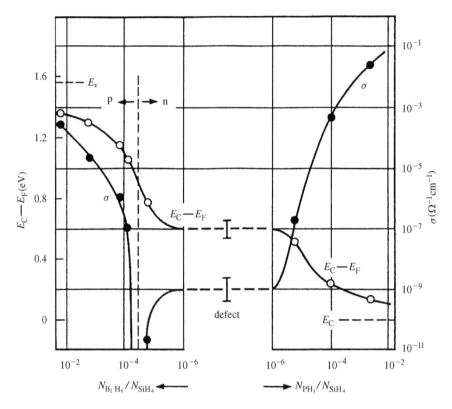

Fig. 1.13 Results of the first pn-control experiments carried out by Spear and LeComber using amorphous silicon [39, reproduced with permission]

substitutional doping as in the case of crystalline Si, with the conductivity σ changed by 10 orders of magnitude on doping with P (phosphorus, group V) or B (boron, group III), and the thermo-electric power switched to n (negative) or to p (positive) as P or B is added, respectively. The thin films prepared in this way contain a considerable amount of hydrogen in the form of Si–H, which plays a key role in the structural sensitivity by disabling any defects. (For this reason, the preparation will be designated as hydrogenated amorphous silicon, a-Si:H, from Chapter 2 onwards.) It is interesting enough that Spear and coworkers failed to notice this fact. That is to say, they did not intentionally prepare hydrogen-containing specimens, but their intuitive insight that the glow discharge technique involved something unusual led to a breakthrough.

Why did Spear stick to the glow discharge process? It is said that Spear replied to this question in the following way: 'Because the higher the photoconductivity is, then the closer the specimen is to an ideal amorphous material'. This is a reasonable answer in view of his background in having studied the photoreceptor surface of a-Se. However, the author (KT) did obtain the following direct answer from Spear himself: 'This was because films prepared by the glow discharge method presented overwhelmingly lower level densities within the gap as measured by the field effect technique in comparison to those made by evaporation or sputtering'. LeComber, the coauthor of Spear's paper, told this author (KT) a more detailed (inside) story as follows: In 1969, Spear and LeComber moved from the University of Leicester to the University of Dundee with their main research theme being 'Drift Mobility and Transport'. They started this work with solid argon (or krypton) because of the drift mobilities being as high as 1000 cm^2 V^{-1}s^{-1}, owing to the lack of optical phonons. Since the mobilities were reduced only by 0.5 even in liquid argon where long-range order is excluded, their interest was naturally directed towards amorphous materials. At this time, the drift mobility of a-Si prepared by evaporation and sputtering was difficult to measure by the time-of-flight method because of the lower resistance, and so they arrived, by elimination, at a-Si grown by the glow discharge technique which had photosensitivity and resistance levels as high as that of an insulator. These three versions are somewhat different from one another in nuance, but they may serve to demonstrate that the use of a-Si grown by the glow discharge technique was an inevitable choice.

It goes without saying that the number of researchers dedicated to amorphous materials has increased dramatically since the publication of Spear and LeComber's work. The 'structure-sensitive' nature is the most important property in view of applications leading to the development of pn junctions, and hence diodes and transistors. In 1976, Carlson and Wronski announced the first report on a solar cell based on hydrogenated amorphous silicon [40], with a number of other studies of applications as thin-film transistors and photoreceptors for xerography starting almost immediately. Further details of these are described in Chapter 6.

1.2.3 ACADEMIC SOCIETIES AND COMMUNITIES

What are the key factors which enable a research topic to expand into a discipline covering an extensive area? The primary factor is, of course, the quality and originality of the research. However, the secondary factor which promotes the growth in an accelerated manner is the absolute number of researchers involved and the presence of relevant academic circles to provide arenas of discussion. I would like to devote some space here to give an historical picture of the research activities in the field of amorphous semiconductors, in a quantitative manner, by reviewing the changes in the related communities, both in Japan and in the rest of the world.

The representative international community in the field of amorphous semiconductors is the *International Conference on Amorphous and Liquid Semiconductors* (ICALS) which holds biennial meetings. The first meeting was held in 1965 in Prague, Czechoslovakia, while the thirteenth was held in 1989 in Asheville, North Carolina, USA. During this period, the 10th meeting of ICALS was held in

Tokyo, Japan in 1983. The first meeting held in Prague was a small symposium organized by Tauc, involving 24 participants from seven European countries, with 17 reports being presented [41]. It should be noted here that Ziman was present among the participants. In the second meeting held in 1967 in Bucharest, Rumania, which was organized by Grigorovici, 40 persons participated, with US scientists joining for the first time [41]. The name ICALS was first adopted at the third meeting held in 1969 in Cambridge, England; this was organized by Mott, with the number of participants increasing to 200, thus taking the form of a genuine international conference. From this occasion onwards, ICALS has since become the base of a large international community involving the vast majority of scientists engaged in studies of amorphous semiconductors. At the 14th meeting, held in 1991 at Garmisch, Germany, the letter 'L' (for liquid) was removed from the acronym for the conference title, so becoming ICAS.

Figure 1.14 presents the annual changes in the number of participants at ICALS (or ICAS) events. This curve shows two-stepped plateaux with the numbers rising rapidly at around 1970 and 1980. The figure also includes plots of the total number of papers involving amorphous semiconductors presented at the two annual meetings of the Japanese Society of Applied Physics (JSAP), namely the spring joint meeting and the fall academic meeting. This curve shows a steep rise following the advent of a-Si:H, clearly indicating the quick response of Japanese researchers to the prospect of possible applications, be this either better or worse! It may be of interest to follow the changes in the session titles in the JSAP meetings: 'Particles and Powders' in 1969, 'Non-Crystalline, Liquids and

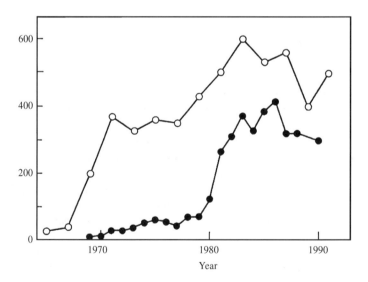

Fig. 1.14   Representations of the annual changes in the number of researchers and papers published in the field of amorphous semiconductors: o, total number of attendants at ICALS; papers reported at the annual meetings of the Japanese Society of Applied Physics

Powders' in 1970, 'Applied Solid-State Physics (Non-Crystalline: Materials and Applications)' in 1978, and 'Non-Crystalline' since the fall meeting of 1985.

Following 1969, the number of publications involving work on amorphous semiconductors submitted to the Japanese Physical Society (JPS) and the Japanese Society of Applied Physics (JSAP) increased at a very slow pace, because there was no forum for making collective discussion. In relation to the JPS, Matsubara, Yonezawa and others (of Kyoto University, at this time) began, around 1970, to plan symposia on non-crystalline semiconductors in connection with the physics of liquid metals, which eventually developed into the Study Group of the Institute of Solid-State Physics, University of Tokyo (organized by Minomura and others) in 1975, and the General Working Group under the Fund of the Ministry of Education (MOE) (organized by Matsubara and others) from 1976.

On the other hand, in relation to the JSAP, a move occurred immediately after the Spring Meeting of 1974, when this author (KT) was contacted by Hirose (of Hiroshima University) who had plans for holding a new seminar. With the research group at the Electrotechnical Laboratory (ETL) as a nucleus and involving Shimizu of Kanazawa University, the plan progressed smoothly, and the first seminar on 'Physics and Application of Amorphous Materials' was held in the fall of 1974 at Kanazawa. This is the origin of the so-called 'Amorphous Seminar' which has been held annually since then, and the first of the intentionally created communities in Japan concerned with amorphous semiconductors. The 'Seminar' was unique in various respects: first of all, there was no one person or organization in charge as the field of study was entirely new. Although the Applied Electron Physics Committee of the JSAP took the core role, this seminar was actually an interdisciplinary complex involving the JPS, the Japanese Institute of Communications Engineers, the Ceramics Association of Japan and various other organizations. Researchers with ages under 35 took active roles in the planning and management of the 'Seminar' and remained at the center of discussion without any territorial conflicts. As the science and technology of amorphous semiconductors were in their very early stages of development, the first participants seemed to have been in a mood of almost 'frontier excitement'.

On the basis of these moves, the 'Amorphous Thin-Film Solar Cell Project' was started in 1980 as one of the priority aims of photovoltaics research in the 'Sunshine Project', a national project sponsored by the Ministry of International Trade and Industry (MITI). This project is characterized by a tripartite collaboration among industrial, academic and governmental sectors and by a broad coverage, including not only the fabrication of solar cells but also basic research (both experimental and theoretical) on the physics and materials science of amorphous solids. The Project is still very active after a period of more than twelve years. In 1980, the Ministry of Education (MOE) embarked on a specific research project, 'Amorphous Materials Science' (organized by Sakurai, Hamakawa and others). After only a few years, the achievements of Japanese researchers were generating high esteem in the international scientific communities, with workers having taken a leading position in several areas. The specific research theme of the MOE had now developed into Committee No. 147, 'Amorphous Materials', of the Japan Society for the Promotion of Science (JSPS).

I should like to emphasize here that Japanese scientists have not only made many excellent contributions to the applications technology of amorphous silicon, but at the same time have also created a number of new and innovative ways of approach to materials science, with the view of achieving structural control and property improvement of materials through physical and chemical studies of the fabrication processes. For further details of these, see Chapter 2 of this book. In the case of a-Si:H, this approach is represented by plasma diagnostics [42]. In just a decade, researchers from many diversified and specialized fields of study, including materials science, dry processing in integrated-circuits technology, nuclear fusion and plasma physics, atomic and molecular spectroscopy, etc., have got together to build up an active community, which subsequently culminated in Committee No. 153, 'Plasma Materials Science', of the JSPS in 1988. Such a case of interdisciplinary union, thus forming a core for a newly emerging field is seldom encountered elsewhere other than in Japan. While this Committee has contributed to raising the technique of plasma CVD analysis to one with the highest international status, this author hopes that Japanese scientists will now turn their attention to the creation of further new and related concepts in this materials research field.

1.3 Organization of the Book

This book, a monograph for specialists concerning the physics and applications of amorphous silicon, is composed of six Chapters and is published by the Japanese Society of Applied Physics as one of the 'Specialists Course' texts in its Applied Physics Series.

Chapter 2 will start with the principles involved in preparing amorphous materials and will then deal with the classification and features of amorphous silicon preparative processes, the principles of the glow discharge process, microscopic diagnosis of the SiH_4 plasma and analysis of the gas-phase processes based upon it, surface growth processes, and modelling of the latter by considering species sorption. In particular, this chapter will describe the mechanisms of the microscopic processes which affect the structure of amorphous silicon, together with the results obtained from plasma diagnosis and a wide range of other analytical techniques.

In Chapter 3, the structural properties of amorphous silicon will be discussed in detail, both at the macroscopic and microscopic levels. With an emphasis placed on the relationship to the preparative processes described in Chapter 2, the factors which determine the various structures of amorphous silicon will be described, including morphology, status of the included hydrogen, degree of structural order, defects, impurities and doping.

In Section 4, the electronic properties of amorphous silicon will be discussed in detail. Starting with the relationship of structural disorder to electronic structure, the description will then cover optical properties, electrical conductivity, the concepts of mobility and mobility edge, doping with impurities, photoconductivity and photoluminescence. This chapter is a particularly important one as it deals with the functions of materials, not only a-Si:H but also alloyed materials such as a-SiGe:H, a-SiC:H and a-SiH:H; multilayered films of a-Si:H with all of these alloys will be described, together with the properties of such materials.

Chapter 5 will be devoted to independent descriptions of structural stability, which are based on the non-equilibrium nature of amorphous materials, with the overall discussion covering thermal stability and photo-induced effects, together with detailed experimental data. In particular, the reversible photoinduced phenomenon, known as the Staebler–Wronski effect, will be treated from all aspects of the present status of research in this area.

Chapter 6 will be concerned with the applications of amorphous silicon. Starting with the various features of amorphous semiconductors, as viewed from the applications side, and their utilization, examples of applications in the many diversified fields will then be described, including solar cells, photosensors, field-effect thin-film transistors and the photoreceptor drum, where comparisons will be drawn with the vidicon-type image pick-up tube using amorphous selenium.

References

1. K. Nishimiya, *Kojiki (Record of Ancient Matters)*, Shincho Series of Japanese Classics, Shincho-sha, Tokyo. (1979), 17 (in Japanese).
2. D. Shechtman, I. Blech, D. Gratias and J. W. Cahn, *Phys. Rev. Lett.*, **53**, (1984) 1951.
3. K. Kimura and S. Takeuchi, *Kotai Buturi (Solid state Phys.)*, **29** (1985) 53 (in Japanese).
4. A. E. Owen, *Electronic and Structural Properties of Amorphous Semiconductors*, edited by P. G. LeComber and J. Mort, Academic Press, London and New York (1973) 161–189.
5. F. Yonezawa, *Amorufasu-na Hanashi (Amorphous Stories)*, Iwanami, Tokyo (1988) 40–57 (in Japanese).
6. J. S. Lannin, *J. Non-Cryst. Solids*, **97/98** (1987) 203.
7. *Amorufasu Handoutai no Kiso (Fundamentals of Amorphous Semiconductors)*, edited by M. Kikuchi and K. Tanaka, OHM, Tokyo (1982) 15–28 (in Japanese).
8. R. M. Schaffert and C. D. Oughton, *J. Opt. Soc. Am.*, **38** (1948) 991.
9. P. K. Weimer, *Phys. Rev.*, **79** (1950) 171.
10. W. E. Spear, *Proc. Phys. Soc.*, **B 68** (1955) 991.
11. N. A. Goryunova and B. T. Kolomiets, *J. Tech. Phys.*, **25** (1955) 984.
12. B. T. Kolomiets, *Phys. Status Solidi*, **7** (1964) 359.
13. A. F. Ioffe and A. R. Regel, *Prog. Semicond.*, **4** (1960) 239.
14. D. Weaire, *Phys. Rev. Lett.*, **26** (1971) 1541.
15. N. F. Mott, *1977 No-beru-sho Koen (Commemorative Speech for Receiving Nobel Prize 1977)* Buturi, 34: 136 (1979) (in Japanese, translated by F. Yonezawa).
16. P. W. Anderson, *Phys. Rev.*, **109** (1958) 1492.
17. P. W. Anderson, *Phys. Rev. Lett.*, **34** (1975) 953.
18. J. L. Hartke, *Phys. Rev.*, **125** (1962) 1177.

19. M. H. Cohen, H. Fritzsche and S. R. Ovshinsky, *Phys. Rev. Lett.*, **22** (1969) 1065.
20. S. R. Ovshinsky, *Phys. Rev. Lett.*, **21** (1968) 1450.
21. A. D. Pearson, W. R. Northover, J. F. Dewald and W. F. Peck Jr, *Advances in Glass Technology*, Plenum Press, New York 1962) 357.
22. M. Kikuchi, *Bussei* (*Solid State Phys.*), **13** (1972) 98 (in Japanese).
23. M. Tanaka and T. Minami: *Yo-Kyo-Shi* (*J. Ceram. Assoc. Jpn*), **72** (1964) 176 (in Japanese).
24. M. Munakata, *Denkishikenjo Hokoku* (*Researches of ETL*), No. 638 (1963) (in Japanese); M. Munakata, *Solid-State Electron.*, **1** (1960) 159.
25. K. Tanaka and M. Kikuchi, *Solid State Commun.*, **13** (1973) 669.
26. T. Igo, Y. Noguchi and H. Nagai, *Appl. Phys. Lett.*, **25** (1974) 193.
27. K. Tanaka: *Oyo Buturi* (*Appl. Phys.*), **47** (1978) 2 (in Japanese).
28. I. Shimizu, H. Sakuma, H. Kokado and E. Inoue, *Bull. Chem. Soc. Jpn*, **46** (1973) 1291.
29. A. Yoshikawa, H. Nagai and Y. Mizushima, Jpn *J. Appl. Phys.*, **16** (1977) Suppl., 67.
30. E. Maruyama, *Sachikon Kaihatsu Shoshi* (*A Short History of Sachicon Development, Hitachi Chuken Shonaiho* (*Bull. Cent. Res. Lab, Hitachi, Ltd*), March 1978–February 1979 (in Japanese).
31. K. Yamamoto, *J. Phys. Soc. Jpn*, **20** (1965) 2291.
32. R. Grigorovici, N. Croitru, A. Devengu and E. Teleman, *Proceedings of the International Conference on Semiconductors, Paris 1964*, Academic Press, New York (1964) 423.
33. J. Tauc, A. Abraham, L. Pajasova, R. Grigorovici and A. Vancu, *Proceedings of the International Conference on the Physics of Non-crystalline Solids, 1965*, North Holland, Amsterdam (1965) 606.
34. R. Grigorovici and R. Manaila, *Thin Solid Films*, **1** (1968) 343.
35. R. C. Chittik, J. M. Alexander and H. F. Sterling, *J. Electrochem. Soc.*, **116** (1969) 77.
36. H. F. Sterling and R. C. G. Swann, *Solid State Electron.*, **8** (1965) 653.
37. A. I. Gubanov, *Sov. Phys. Solid State*, **3** (1962) 1694.
38. N. F. Mott, *Phil. Mag.*, **19** (1969) 835.
39. W. E. Spear and P. G. LeComber, *Solid State Commun.*, **17** (1975) 1193.
40. D. E. Carlson and C. R. Wronski, *Appl. Phys. Lett.*, **28** (1976) 671.
41. R. Grigorovici, *J. Non-Cryst. Solids*, **97/98** (1987) 1463.
42. K. Tanaka, *Oyo Buturi* (*Appl. Phys.*), **60** (1991) 361 (in Japanese).

2 METHODS OF PREPARATION AND GROWTH PROCESSES

In this chapter, we will review the preparation and growth processes of amorphous silicon. Materials in a thermodynamic non-equilibrium state may take various structures depending upon the preparative process. It is vital when carrying out research on amorphous materials to observe and understand the microscopic processes of structure formation and to be able to achieve accurate control over the latter. Starting with the reactor construction, this survey will cover the physics and chemistry of glow discharge, the primary and secondary processes within the SiH_4 plasma, and analysis of the steady-state conditions through various methods of plasma diagnostics. A detailed discussion of the growth processes will then follow, based upon experiments and theory concerning reactions occurring at the surface of the material.

2.1 Outline of Preparative Methods

2.1.1 NON-EQUILIBRIUM PROCESSES AND AMORPHOUS STRUCTURES

Thermodynamically speaking, amorphous materials may be regarded as solids in a non-equilibrium state (see Section 1.1.2). Accordingly, the preparation of amorphous solids may be regarded as a technique for implementing a non-equilibrium state, in which the state is always frozen through a non-equilibrium state. The preparation of amorphous solids may be classified in terms of the initial phases as follows: (a) the freezing of a liquid, (b) the freezing of a gas and (c) the modification of a solid. The corresponding non-equilibrium processes for these are: (a) rapid quenching; (b) gas condensation or chemical reactions/physical processes at the gas–solid interface and (c) ion implantation. The relationship between these processes is illustrated in Fig. 2.1 [1]. While techniques (a) and (b) are frequently used for the preparation of amorphous semiconductors, technique (c) is not commonly used. All materials can be made amorphous by using the gas freezing process (b).

The difficulty in forming an amorphous structure varies widely depending upon the material itself. An amorphous semiconductor consists of a network largely

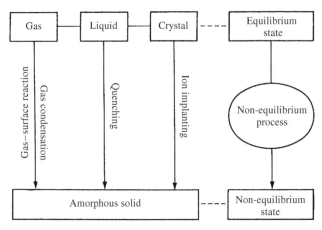

Fig. 2.1 Relationship between the various preparative methods used to produce amorphous solids [1, reproduced with permission]

built up by using covalent bonds, with the mechanical stability of this structure being closely correlated with the coordination numbers of the various atoms. As stated in Section 1.1.4, the coordination numbers of the main component elements in solids of chalcogenides, pnictides and tetrahedrally bonded groups are 2 (Se), 3 (P) and 4 (Si), respectively, with the network features changing from one- to three-dimensional, according to Mott's empirical rule, i.e. the '8-N' rule, where N is the number of electrons in the outermost shell. Phillips calculated the number of constraints per atom, N_c, on the basis of the mean coordination number of the constituent atoms, M, as follows [2]:

$$N_c = \frac{M}{2} + \frac{M(M-1)}{2} = \frac{M^2}{2} \tag{2.1}$$

The mean coordination number M of the amorphous semiconductor $A_x B_y$ is defined by using the coordination numbers, $M(A)$ and $M(B)$, and the atomic ratios, x and y, of the elements A and B, respectively, where $x + y = 1$:

$$M = xM(A) + yM(B) \tag{2.2}$$

The first term of equation (2.1) represents the degree of freedom for bond stretching, while the second term represents that of bond bending. Based on the assumption that the network is in its most stable form when N_c is equal to the number of dimensions in normal space, i.e. 3, Phillips calculated M for $N_c = 3$, as follows:

$$M = 2.45 \tag{2.3}$$

(This value was later modified by Döhler and coworkers who obtained $M = 2.4$ [3].)

In cases where the coordination number (the number of bonds per atom) is much greater than 2.4, the number of constraints N_c increases to retain more strain energy (enthalpy) in the amorphous structure network. If M is much less than 2.4, it may happen that the configurational entropy increases, thus reducing the glass transition temperature, i.e. the liquid phase is the most likely state, or that the enthalpy becomes more difficult to freeze and thus the system is less likely to crystallize, even if kept at lower temperatures. Hence, as shown in Fig. 2.2, a representative chalcogenide, a-As$_2$S$_3$, which has $M(As) = 3$, $M(S) = 2$, $x = 0.4$ and $y = 0.6$, resulting in $M = 2.4$ from equation (2.2), may be regarded as having a mechanically most stable amorphous structure. In fact, a-As$_2$S$_3$ readily turns into a glass, even when slowly cooled from the liquid phase, and it is actually rather difficult to obtain the crystalline state. On the other hand, it is impossible to prepare a-Si ($M = 4$) by quenching from the liquid phase. In the case of hydrogenated amorphous silicon (a-Si:H), the presence of hydrogen effectively reduces M, thus relieving the strain energy of the network. It is generally known that if $M > 3$, an amorphous solid can only be obtained by freezing from the gas phase.

The analysis by Phillips mentioned above is simple and involves the essential points, although it may be inadequate for evaluating the difficulty of forming amorphous structures for actual materials. For example, the ionicity of the chemical bonds must be taken into consideration. In the case of a fully covalent bond, with an ionicity equal to zero, the directionality of the bond is too strong to produce an amorphous structure, and in the case of complete ionic bonding, the bond-bending force constant of the two bonds becomes zero (i.e. the concept of a bond essentially disappears) and thus fails to produce an amorphous structure. That is to say, in order to implement an amorphous structure, covalent bonds with a certain degree of ionicity are needed.

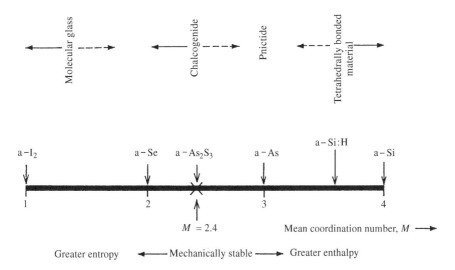

Fig. 2.2 Classification and mechanical stability of amorphous semiconductors based on the mean coordination number M [1, reproduced with permission]

I would like now to discuss in detail two of the major methods which are used to prepare amorphous solids.

(a) Liquid quenching method

As shown in Fig. 1.1 in Section 1.1.2, this method is used to prepare amorphous solids via the supercooled liquid phase, with a bulk amorphous solid (glass) being obtained. This technique is effective only for chalcogenide and pnictide materials, and is not applicable to tetrahedrally bonded materials. It has been reported that the surface layer of a silicon crystal is made amorphous by irradiation with a high-power, ultra-short-pulsed laser. However, this is a rather unusual method of preparation, and it should be noted that the starting material is not liquid (with a coordination number of 6 for Si) but solid [4].

Since both chalcogenide and pnictide materials contain elements which have high vapour pressures or are readily oxidizable, they are very difficult to prepare by simple procedures such as those used for oxide glasses [5]. Generally, highly purified and accurately weighed amounts of the elements are sealed in fused quartz containers in a vacuum and then heated to high temperatures for some 10 hours or more in a rocking furnace (thus both melting and stiming), and then quickly cooled to room temperature. The quenching rate can be varied widely depending upon the cooling method being used, e.g. either by leaving in the furnace or by putting into water. In the case of multi-component glasses, the higher the quenching rate, then the broader will be the composition range of glass formation [6]. (see Chapter 1).

In the splat cooling method, which is often used for preparing amorphous metals, and which involves splatting a small amount of melt to a bulk metal, the quenching rate is said to be as high as 10^4 °C s^{-1} [7]. For the mass production of thin sheets (several tens of microns in thickness) of amorphous metal, both the single-roll method which feeds a melt continuously and steadily on to a roller rotating at high speed, or the double-roll method which sweezes a melt between two rollers, are well known, but neither of these techniques are applicable to the preparation of amorphous semiconductors [7].

(b) Gas condensation method

Amorphous semiconductor thin films of chalcogenide or pnictide materials are prepared by evaporation or sputtering, starting from the respective bulk glass systems prepared by the liquid annealing process. These methods are collectively known as physical vapour deposition (PVD) techniques.

As an example, in the case of the evaporation of a-As_2S_3, bulk As_2S_3 is pulverized and heated to evaporate at as low a temperature as possible and a film is then grown on a substrate which is kept at low temperatures. Flash evaporation at higher temperatures causes revaporization of S to change the composition, thus resulting in As S film containing excess As [6]. It has been demonstrated by using Raman scattering and X-ray diffraction analysis that as-evaporated a-As_2S_3 thin films closely resemble small-molecular aggregates such as As_4S_4 and As_4S_6 [6]. For building up a continuous network such as a bulk glass, heat treatment at appropriate temperatures (110–120°C for a-As_2S_3) is indispensable. In the case of the sputtering process, in which no particular molecular state exists, heat treatment is not required.

On the other hand, in the case of tetrahedrally bonded materials, the PVD process is applicable to the preparation of a-Si and a-Ge by using their respective crystalline forms as starting materials. However, the a-Si:H discussed in this present book can be prepared only through chemical vapour deposition (CVD) techniques such as glow discharge, reactive sputtering and photochemical vapour deposition. The major difference between CVD and PVD is that the gas pressure within the reaction chamber for the former is set so that the mean free path of the particles in the gas phase is far smaller than the geometrical dimensions of the chamber in order to promote reactions in the gas phase.

2.1.2 PREPARATION OF AMORPHOUS SILICON

The various methods for preparing a-Si:H are shown in Table 2.1. Among these methods, the most intensively studied is the plasma CVD method [8], followed by the reactive sputtering method [9], the thermal CVD method [8] and the photo CVD method [10]. The preparation of a-Si:H by using these different methods will be described below.

(a) Glow discharge (plasma CVD) method

The glow discharge method is a plasma CVD process in which SiH_4 (often diluted with H_2 or Ar) is decomposed in a glow discharge and the Si is then deposited on to a substrate held at 200–300°C [8]. Samples of a-Si:H prepared by this method contain 10–20 atom% of hydrogen. The reactor design is usually of two types, namely the capacitively coupled (diode) type and the inductively coupled (electrodeless) type as shown in Fig. 2.3. While a radiofrequency (RF) plasma of 13.56 MHz is normally used, direct current (DC) glow discharge can also be used for the former. The glow discharge is characterized by the use of a non-equilibrium plasma (low-temperature plasma), which forms a non-equilibrium gas phase which is substantially different from the usual thermal CVD method. 'Non-equilibrium' used here means that the electron energy in a plasma is 1–10 eV, corresponding to 10^4–10^5 K in the electron temperature, T_e, the translational temperature, T_{gas}, of the gas molecules (or atoms) is close to the temperature of the apparatus (the normal temperature), i.e. $T_e/T_{gas} \cong 10^2$–10^3, and such a high thermal energy or temperature as to decompose the molecules is not needed. The kinetic energy of the electron is responsible for the decomposition of the molecules. By collision with an electron, a molecule is activated to a higher energy state, and its dissociation or ionization then takes place. For this reason, the glow-discharge is called a low-temperature plasma. If B_2H_6 or PH_3 is mixed with the source gas SiH_4 in the glow discharge process, B (boron) or

Table 2.1 The various processes and conditions used for preparing amorphous silicon (a-Si:H)

Method	Source	Growth conditions	Reference
Plasma CVD method			8
DC glow discharge	$SiH_4(H_2)$	10^{-2}–1 torr, $T_S = 100$–300°C	28
RF glow discharge (1–13.56 MHz)	$SiH_4(H_2, Ar)$	10^{-2}–1 torr, $T_S = 100$–300°C	39
ECR[a] method (GHz band)	$SiH_4(H_2)$	10^{-5}–10^{-3} torr, $T_S = 100$–300°C	10
Remote plasma method	$SiH_4(H_2)$	10^{-2}–1 torr, $T_S = 100$–300°C, H_2 separated from SiH_4	11
Reactive sputtering method	Crystalline Si target, $(Ar + H_2)$ mixed gas	10^{-5}–10^{-2} torr, $T_S = 100$–300°C	8, 9, 15, 16
Photo CVD method			23, 25
Direct excitation method	Si_2H_6	$h\nu = 2537$ Å, 1849 Å, $T_S = 100$–300°C	24
Mercury sensitization method	SiH_4, Hg	$h\nu = 2537$ Å, $T_S = 100$–300°C	26
Thermal CVD method			17
Thermal CVD method	$SiH_4(H_2, Ar)$, Si_2H_6	Normal pressure, $T_S = 500$–650°C (SiH_4)	17, 19
Homogeneous CVD method	SiH_4	Normal pressure, $T_S = 150$–350°C	18
Excited-species CVD method	SiH_4, Ar	Ar excited by plasma and fed into SiH_4	29
ICB[b] method	Crystalline Si, H_2	10^{-7}–10^{-4} torr, Si from crucible to ICB[b]	30
Liquid-reaction method	$SiHCl_3$	No systematic data	33

[a]ECR: electron cyclotron resonance
[b]ICB: ion-cluster beam

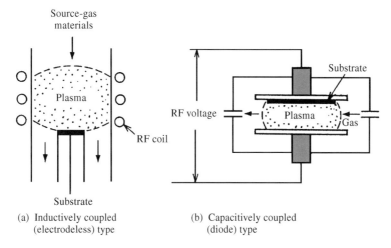

(a) Inductively coupled
     (electrodeless) type

(b) Capacitively coupled
     (diode) type

Fig. 2.3 Schematic representations of the two types of reactors used in the glow discharge method

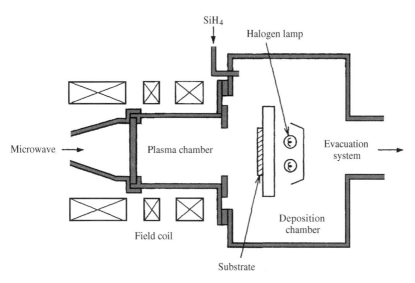

Fig. 2.4 A schematic representation of a typical ECR plasma CVD system [10, reproduced with permission]

P (phosphorus) can be easily doped into a-Si:H, without requiring any diffusion of B or P at higher temperatures. In addition, there are other methods available, such as the ECR plasma method [10], which uses a frequency of 2.45 GHz, and the remote plasma method [11] where SiH_4 is excited at a different location from the Ar or H_2. In these ways, high-quality films of a-Si:H can be prepared via several varieties of plasma CVD processes. Fig. 2.4 shows the construction of the ECR plasma apparatus, which is characterized by high energy efficiency as a result of the cyclotron resonance of the electrons, which permits films to be grown under higher vacuum conditions (10^{-5} torr). This method is of particular merit for producing alloy-based films such as a-Si_xC_y:H.

(b) Reactive sputtering method

In this technique, using crystalline Si as a target and a mixture of argon and hydrogen as the sputtering gas, a-Si:H can be deposited on substrates kept at 200–300°C. This method is characteristically capable of controlling the hydrogen content in a-Si:H with great accuracy, and without any constraints, by changing the ratio of Ar to H_2 [8, 9]. Since the electron temperature of the plasma is several times higher than that used in the glow discharge (plasma CVD) method, all of the atoms and molecules are readily activated, and impurities are thus likely to be included in the a-Si:H. Under certain conditions, a few atom% of Ar may also be included [12]. It is rather difficult to control the doping dosage exactly, but the mechanical adhesion of the a-Si:H to the glass substrate is found to be very good.

The sputtering technique was the main method of a-Si preparation before 1975, i.e. prior to the successful pn control of a-Si:H, and has been intensively studied with respect to the structure and electronic properties of the resulting materials. A structural model containing voids of 5–10 Å in diameter is well established [13]. Paul and his group at Harvard University prepared a-Ge:H through the reactive sputtering method in 1974 [14], and demonstrated that the defect density can be considerably reduced by introducing hydrogen into a-Ge or a-Si [15]. Therefore, it was experimentally proved for the first time by using the reactive sputtering method that hydrogen plays a vital role in reducing defects. As more recent work has achieved major improvements with respect to film quality and other factors, this method is likely to achieve prominence yet again [16].

(c) Thermal CVD method

The thermal CVD method is a process in which SiH_4 is thermally dissociated at temperatures of 500–650°C with a-Si films being deposited on a substrate held at similar temperatures [17]. As shown in Fig. 2.5, SiH_4 is diluted with H_2 or Ar before being fed into the reaction tube. The activation energy for optimum growth rate is 1.5 eV (34.6 kcal mol^{-1}), and if the temperature exceeds 680°C, the film becomes polycrystalline. If doped with P, the activation energy for optimum growth increases beyond 1.5 eV, while when doped with B, it is considerably reduced [17]. This method is simple, and is very suitable for mass production of material but it does have certain disadvantages, such as requiring higher temperatures than other processes, and can lead to products containing higher densities of defects, owing to the inadequate hydrogen content of the films. In order to overcome these problems, attempts have been made (1) to grow films with the substrate temperature reduced to below 350°C (homogeneous CVD) [18], (2) to

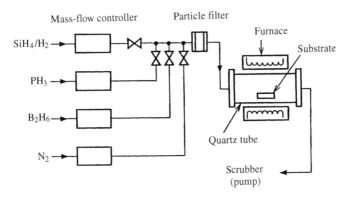

Mass-flow controller   Particle filter

Furnace

SiH$_4$/H$_2$ →

Substrate

PH$_3$ →

B$_2$H$_6$ →

Quartz tube

N$_2$ →

Scrubber
(pump)

Fig. 2.5 A schematic representation of a thermal CVD system used for preparing and doping a-Si [17, reproduced with permission]

use higher-order silane source materials, such as Si$_2$H$_6$ or Si$_3$H$_8$ (which are dissociated at lower temperatures), in place of monosilane (SiH$_4$) [19], and (3) to treat the as-grown film with a hydrogen plasma in order to introduce further hydrogen into the film (post-hydrogenation technique) [20, 21]. The doping properties of a-Si prepared by this method are generally inferior to those of a-Si prepared by the glow discharge method [22].

(d) Photo CVD method

In contrast to the glow discharge method in which SiH$_4$ is activated, dissociated and decomposed by using the kinetic energy of electrons, another type of process, the photo CVD method, makes use of photon energies to decompose the SiH$_4$, either directly or indirectly. As in the case of the glow discharge method, the process may be operated at lower temperatures (150–300°C). In comparison to plasma excitation, in which the excitation is induced by collisions with electrons, photo CVD has the following benefits: (1) the selectivity of excitation is very high because multiple forms of the molecular excitation are available depending upon the wavelength of the radiation, and (2) films can be deposited over spatially selected areas by the use of lasers [23]. This is in striking contrast to the excitation induced by electron collision, which does not follow the selection rules for optical transitions, i.e. it is readily excitable, but not widely selectable.

There are two types of photo CVD methods used in the case of SiH$_4$. The first of these directly dissociate the SiH$_4$ by using radiation with wavelengths within the absorption band of the silane (the direct photo CVD method) [24]. In this method, the reaction always occurs via a singlet excitation state [25], as shown in Fig. 2.6, and the photon energy, $h\nu$, must be $\geqslant 8.2$ eV (corresponding to UV irradiation of wavelength <153 nm). The second technique is the mercury sensitization (indirect photo CVD) method [26]. In this, mercury is excited to the triplet state, and its energy is then transferred to the silane without passing through the

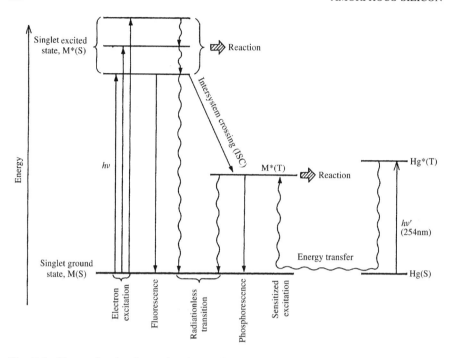

Fig. 2.6 Energy levels of a molecule (M) involved in the photochemical processes and sensitization by mercury (photo CVD methods), where S represents the singlet and T the triplet states. For heavy metals, such as mercury, the selection rule no longer holds [25, reproduced with permission]

singlet state, thus allowing the use of radiation of longer wavelengths for excitation purposes (see Fig. 2.6) [25]. If Si_2H_6 or Si_3H_8 are used for source materials in place of SiH_4, the absorption band is shifted toward the longer-wavelength side [27], and it then becomes possible to use radiation of longer wavelengths for excitation even in the direct photo CVD method [24]. This technique is expected to be widely exploited in the future.

(e) Other methods

In addition to the methods discussed above, Table 2.1 lists a number of other methods which are used to fabricate amorphous silicon, including the DC glow discharge method [28], the excited-species CVD method [29], and the ion-cluster-beam (ICB) method [30]. For further details of these, please refer to the papers cited. In addition, the hot-wire CVD method [31] and the chemical-deposition method [32] have also been reported. The liquid reaction method (electrolytic deposition) is a technique of a somewhat different nature [33]. Starting with $SiHCl_3$, and using either $AsCl_3$, $GaCl_2$, or BBr_3 as dopants, films are deposited electrolytically, and later subjected to heat treatment at 350–450°C.

It has been reported that a-Si:H films of fairly good quality can be obtained by this method, but the process has not been systematically studied in great detail.

2.2 Formation of SiH$_4$ Plasma

2.2.1 PLASMA REACTOR CONSTRUCTION AND ELECTRON ENERGY DISTRIBUTION

Before starting the discussion on the decomposition processes of SiH$_4$ in a plasma, let us consider some of the basic facts concerning RF plasmas as used in the capacitively coupled (diode) type reactor for the glow discharge (plasma CVD) method described in Section 2.1.2 (a).

Figure 2.7 shows schematically the basic construction of an RF glow discharge reactor and the electric field distribution across the plasma between the two electrodes [34]. When the gas pressure is set at $10^{-2}-1$ torr, and an RF voltage of $10-100$ V (13.56 MHz) is applied between the RF-side electrode (area S_1) and the (electrically grounded) substrate-side electrode (area S_2, which includes the contribution from the grounded reactor wall), SiH$_4$ is dissociated and ionized by collision with electrons thus generating a sustained glow discharge plasma. The plasma positive column shown in Fig. 2.7 represents the glow discharge, emitting light which is normally composed of bright-line spectra (see below). The potential in this region, known as the plasma potential (V_p), is always positive with respect to the ground potential, and its level is at least as high as the primary ionization potential of the gaseous materials (atoms or molecules). It should be noted that the substrate installed on the grounded electrode is negative with

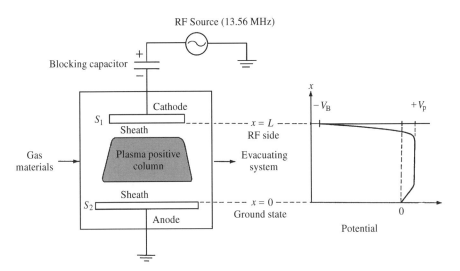

Fig. 2.7 Schematic representation of the basic construction of an RF glow discharge reactor, showing the potential distribution between the electrodes in steady-state plasma operation

respect to the plasma, and is therefore constantly exposed to bombardment by positive ions.

On the other hand, the RF side electrode is always negative with respect to the plasma, because of the difference in mobility (under the electric field) between the electrons and ions (mainly cations) within the plasma, i.e. μ_e and μ_i, respectively. Owing to the difference in mass between an electron and an atom, the mobility of the electron is much greater than that of the corresponding cation:

$$\mu_e \gg \mu_i \tag{2.4}$$

Under an RF field in the MHz region, only the electrons can follow the polarity change in the field, while cations fail to oscillate. Consequently, a negative charge is accumulated on the blocking capacitor connected to the RF side electrode shown in Fig. 2.7 (for this reason this electrode is called the cathode), and a DC voltage (self-biased, $-V_B$) appears. This takes place in the non-emitting area known as the plasma sheath [34].

The frequency at which the ions fail to respond to the changing field, namely the cut-off frequency f_c, is given by the following expression:

$$f_c = \frac{eE\lambda_i}{2\pi m_i v_i L} \tag{2.5}$$

where m_i is the mass of the ion, v_i the ion velocity, λ_i the mean free path, e the charge of the electron, L the distance between the anode and the cathode, and E the amplitude of the RF field. If $E = 100$ V cm^{-1} and $L = 5$ cm, then f_c is within the range 1–100 kHz [35].

The magnitude of V_B varies depending upon the ratio of the areas of the top and bottom electrodes, the ion mass and the gas pressure. The ratio of the voltage drop at the plasma sheaths for the two electrodes is a function of the effective electrode area ratio, as follows:

$$\frac{V_p}{V_p + V_B} = \left(\frac{S_1}{S_2}\right)^\alpha \tag{2.6}$$

where α is ca 4 (theoretically) or ca 1 (experimentally) [34], i.e. if the relative electrode area is reduced, then the voltage drop at the electrode rises steeply.

Within the plasma positive column, electrons accelerated by the voltage V_p collide with molecules to either dissociate or ionize them. In the vicinity of the cathode (the RF side) where the self-biased high voltage ($-V_B$) is effective, as illustrated in Figure 2.7, electrons of high energy accumulate. For this reason, the most intensive emission occurs around the cathode in the positive column, and the rate of molecular dissociation, and hence, the rate of SiH$_4$ consumption, is at a maximum [36] (see also Fig. 2.12 below).

Exact data on the distribution of the electron kinetic energy $f(E)$ in the RF glow discharge of SiH$_4$ are not available because of experimental difficulties. It is known, however, that due to the ionization and dissociation of SiH$_4$, the distribution of the electron energy deviates markedly from the Maxwell distribution. As a reference, Kocian's measurement of the energy distribution in the

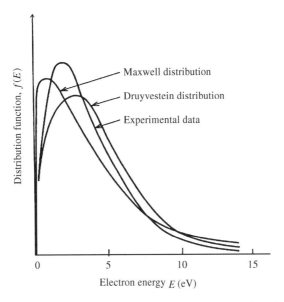

Fig. 2.8 Electron energy distributions in a SiH_4 plasma [37, reproduced with permission]

DC discharge of SiH_4 is shown in Fig. 2.8 [37]. As will be described below, electrons only having energies higher than 8 eV are used for dissociating SiH_4. The mean energy of these electrons corresponds to an electron temperature T_e of *ca* 10^4 K, which is two orders of magnitude higher than the gas temperature, T_g. Because of this, the plasma is designated as being non-equilibrium in nature, as stated earlier.

2.2.2 ELECTRON IMPACT DISSOCIATION OF SiH_4

The overall dissociation rate constant k_T for electrons with an energy distribution $f(E)$ colliding with the source molecules of SiH_4 to generate various neutral radicals and ions is given as follows:

$$k_T = \int_0^\infty f(E)\sigma_T(E) \left(\frac{2E}{m_e} \right)^{1/2} dE \qquad (2.7)$$

where m_e is the mass of the electron and $\sigma_T(E)$ is the total cross-section for dissociation and ionization of SiH_4 with respect to an electron with energy E. Schmitt and coworkers succeeded, by monochromating the electron energy and reducing the gas pressure ($\sim 10^{-4}$ torr), in experimentally determining the dependence of the dissociation cross-section on the electron energy [38]. Fig. 2.9 shows this dependence for SiH_4 for dissociation, ionization and species generation caused by inelastic collision with the electrons.

As is evident from this figure, the cross-section σ_T for the electron-impact dissociation of SiH_4 has a threshold at around $8-9$ eV, while the cross-section

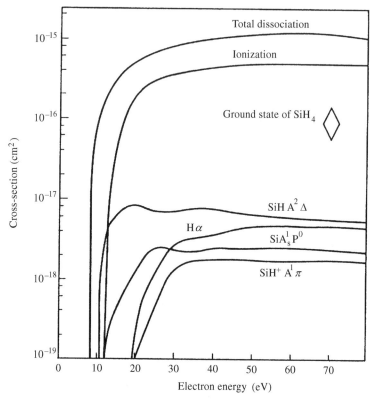

Fig. 2.9 Dependence of the electron collision cross-section of SiH_4 on electron energy, shown for total dissociation, ionization and generation of various species [38, reproduced with permission]

for ionization σ_{ion} is a little higher, at about 11.9 eV. The four lower curves shown in Fig. 2.9 represent the electron collision cross-section data for generating atoms and molecules in various excited states, as obtained from the line emission spectra. While no data are available for the generation of neutral radicals and atoms in the ground state at the present time, the pathways for both the primary electron impact decomposition of SiH_4 and radical generation can be represented by the scheme shown in Fig. 2.10 [39]. On the basis of various experiments, it may be claimed that for electrons of distribution $f(E)$, as shown in Fig. 2.8, SiH_2 and SiH_3 are generated with the highest efficiency, followed by SiH and Si, in that order. The cross-sections for the generation of ions and excited species have much lower values than these, as shown in Fig. 2.9. It should be noted that the cross-section for ionization for electron energies of 8–13 eV is very small, as shown in Fig. 2.9 [38, 39]. The dissociation process of SiH_4 caused by electron impact, and hence, the initial process of plasma reaction is as explained so far.

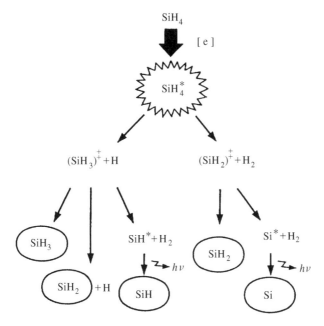

Fig. 2.10 Illustration of the primary decomposition processes of SiH$_4$ by electron collision and generation of neutral radicals [39, reproduced with permission]

2.2.3 STEADY-STATE CONDITIONS OF THE SiH$_4$ PLASMA: MERITS OF PLASMA DIAGNOSIS

After dissociation into various nuclear species brought about by the inelastic collision with electrons, diversified reactions may occur within the plasma as secondary and tertiary processes, involving generated species, parent molecules of SiH$_4$ and electrons. Providing that parameters such as gas flow, pressure, and input power to the plasma are kept constant, the concentrations of the various species should reach steady-state conditions in the reactor.

The concentration of the ith species, N_i, in the plasma (including neutral molecules, radicals, neutral atoms and ions) can be described by the following partial differential equation [40]:

$$\frac{\partial N_i}{\partial t} = \nabla[D_i \nabla N_i] - n k_i N_i - \sum_j k_{ij} N_j N_i + \sum_{jk} k_{ijk} N_j N_k + \sum_j n k'_{ij} N_j \quad (2.8)$$

where D_i is the diffusion coefficient of the ith species, n the electron density, k_i the overall rate constant for dissociation, ionization and electron adhesion of the ith species on collision with electrons, N_i and N_k the concentrations of the jth and kth species, respectively, k_{ij} the rate constant for the chemical reaction between the ith and jth species, k_{ijk} the rate constant for the chemical reaction which generates the ith species from the jth and kth species, and k'_{ij} the rate constant for the reaction which generates the ith species by collision of the jth species with an electron. Furthermore, it is assumed that the mean free path of the various species is far smaller than the reactor dimensions, and that the diffusion time within the reactor is far shorter than the residence time of each molecule. (This is a fairly realistic assumption for the preparation of a-Si:H.) The rate constant k_i in equation (2.8) is equivalent to k_T in equation (2.7), if the ith species is SiH_4.

These rate constants are functions of both the electron energy and temperature, while the diffusion coefficient depends upon the temperature, concentration and gas composition. If all of these parameters are known, and boundary, as well as initial conditions, such as the supply and removal of source gas, and adsorption and desorption processes at the substrate, electrode surface and wall surface are given, then the behavior of each species can be described by solving equation (2.8). In particular, the concentration distribution in the steady state is obtained from $(\partial N_i / \partial t) = 0$.

In the actual system, however, systematic fundamental data for electron impact cross-section and chemical reaction rate constants are too insufficient to solve exactly equation (2.8). Several studies have been reported on the concentration of species determined by computer simulation experiments on the basis of various assumptions and approximations [41]. On the experimental side, plasma diagnostic data accumulated over some 10 years since 1978 have provided marked contributions [39, 40, 42]. The concentrations of neutral molecules, radicals, excited atoms and ions in an RF SiH_4 plasma in the steady state have been fully determined by various plasma diagnostic methods. Under the conditions of a SiH_4 flow rate of 1 ml s^{-1}, a pressure of 30 m torr, and a reactor volume of 3×10^4 cm^3, the concentrations of the various species in the SiH_4 plasma have been reported as follows: $[SiH_4] = 10^{15}$ cm^{-3}, $[SiH_3] = 10^{12}$ cm^{-3}, $[SiH_2] = 10^{10}$ cm^{-3} (or less), $[SiH_x{}^+] = 10^9$ cm^{-3} (the concentration of the cations are equal to the sum of the electron concentrations and the anion concentrations), and $[Si^*]$ and $[SiH^*] = 10^6$ cm^{-3} [39, 43, 44].

Let us try to estimate from equation (2.8) the level of the concentration of SiH_3, which is regarded as being relatively stable in a SiH_4 plasma (to be discussed further below). It is assumed that in equation (2.8), the major terms are the following:

$$N_i = [SiH_3]$$

$$\nabla[D_i \nabla N_i] = -D \left(\frac{L}{2}\right)^{-2} [SiH_3]$$

$$\sum_j n k'_{ij} N_j = n k_T [SiH_4]$$

and all other terms can be ignored. Therefore for the following expression:

$$\frac{\partial [SiH_3]}{\partial t} = -D \left(\frac{L}{2}\right)^{-2} [SiH_3] + n k_T [SiH_4] \qquad (2.9)$$

Table 2.2 Plasma diagnostic methods used for SiH_4 RF plasma experiments and their characteristic features

Diagnostic method	Nuclear species to be measured	Level of detection	Spatial resolution	Remarks
IR diode laser absorption spectroscopy (IRLAS) [45]	Neutral molecular species (ground state), SiH_4, SiH_3, SiH_2, SiH	$10^9 - 10^{10}$ cm^{-3}	Good	Excellent detection limit
Coherent antistokes Raman spectroscopy (CARS) [40]	Neutral molecular species (ground state), SiH_4, SiH_2, H_2, Si_2H_6	10^{13} cm^{-3}	Excellent	Gas temperature measurable
Laser-induced fluorescence method (LIF)	Light-emitting species (ground state), SiH_2, SiH, Si_2	10^6 cm^{-3}	Excellent	Quantitative data not available
Optical emission spectroscopy (OES) [44]	Light-emitting species (excited state), SiH^*, Si^*, H_2^*, H^*	10^6 cm^{-3}	Good	Quantitative data not available, simple method
Ion Mass Spectrometry [44] [50]	Ions (cations)	—	Poor	Neutral species not measurable
Langmuir-probe method	Electron density, electron energy	—	Good	Plasma disturbed, probe stained

if we put $\partial[SiH_3]/\partial t \equiv 0$, $[SiH_4] = 10^{15}$ cm^{-3}, $D = 5 \times 10^3$ cm^2 s^{-1}, $L = 5$ cm, $n = 10^8$ cm^{-3} (electrons having energies of 8 eV or higher are assumed to have about one tenth of the electron density value of 10^9 cm^{-3}), and $k_T (= v_e \sigma_T) = 10^{-8}$ cm^3 s^{-1} (most of the dissociated SiH$_4$ is assumed to be SiH$_3$), we obtain $[SiH_3] \cong 10^{12}$ cm^{-3}. This value agrees well with the experimental data obtained from plasma diagnostics.

Table 2.2 summarizes the various plasma diagnostic methods used for SiH$_4$ RF plasma experiments and their characteristic features. Each of these methods, with the exception of the classical Langmuir-probe method, will be described individually below.

(a) Infrared (diode) laser absorption spectroscopy (IRLAS)

Figure 2.11 shows the measuring system used for infrared (diode) laser absorption spectroscopy (IRLAS) involving a multiple reflection system [45]. Instrumentation, based on the infrared diode laser and state-of-the-art electronics has recently become available. The concentrations of SiH$_3$, SiH$_2$ and SiH radicals in their ground states are assessed by measuring the infrared absorptions for the respective vibrational–rotational transitions. The detection limit can be as high as 10^9–10^{10} cm^{-3}. The concentration of the SiH$_3$ radical, which is the most important radical in the deposition of a-Si:H, was determined by this method to be 10^{11}–10^{12} cm^{-3} [43]. The steady-state concentration [SiH$_3$] distribution between the reactor electrodes was measured as shown in Fig. 2.12 under actual RF glow discharge conditions [43].

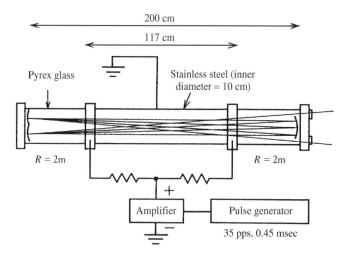

Fig. 2.11 Schematic representation of the measuring system used for infrared diode laser absorption spectroscopy (IRLAS) which utilizes a multiple reflection system [45, reproduced with permission]

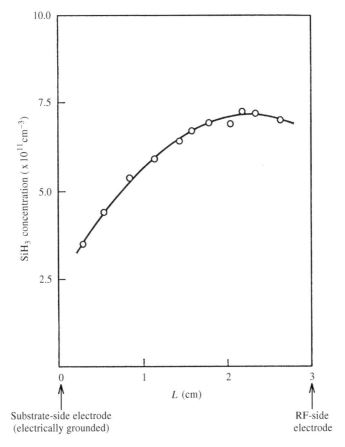

Fig. 2.12 Concentration of SiH_3 as a function of the distance between the reactor electrodes measured under RF glow discharge conditions [43, reproduced with permission]

(b) Coherent anti-Stokes Raman spectroscopy (CARS)

This method utilizes non-linear Raman scattering in place of infrared absorption. Laser beams of two different frequencies, ω_1, and $\omega_2(\omega_1 > \omega_2)$, overlaid coaxially are passed through the reaction gas (SiH_4 plasma). When the vibrational–rotational transition energy, $\Delta\omega$, of the molecular species in the ground state contained in the gas phase achieves the condition, $\Delta\omega = \omega_1 - \omega_2$, a coherent anti-Stokes spectral signal of frequency, $\omega_3 = 2\omega_1 - \omega_2$, is generated. Accordingly, if ω_1 is fixed and ω_2 is varied so as to scan over a certain range, a discrete spectrum of ω_3 is obtained, thus allowing the identification of the molecular species and the determination of their concentrations on the basis of wavelength and intensity [36, 40]. The merits of this particular method are excellent spatial resolution and a capability of determining the temperature of the molecular species on the basis of their rotational energy spectra [40]. Data for the concentrations of SiH_4, Si_2H_6, SiH_2 and H_2 have been published. The detection limit is around 10^{13} cm^{-3}.

(c) Laser-induced fluorescence (LIF) method

In the laser-induced fluorescence (LIF) method, neutral molecular species (in the ground state) are excited to a higher electronic state by using a laser beam; the wavelength and intensity of the fluorescence emitted when the species are deactivated to the ground state via a light-emitting level are then measured. When the detection limit is 10^6 cm^{-3} or so, the quantitative determination of concentrations can involve some problems. Data for SiH_2, SiH, Si and Si_2 have been reported [46].

(d) Optical emission spectroscopy (OES)

In this method, the light emission from a plasma is directly analyzed spectroscopically, and the various species in their excited states are identified from the line emissions. It may be claimed that SiH_4 plasma diagnosis was largely initiated by application of this spectroscopic technique [44, 47, 48]. Fig. 2.13 shows a typical optical emission spectrum of a SiH_4 plasma. It is known empirically that the line intensity of Si^* (2 881 Å: UV 43) and SiH^* (4 127 Å: $A^2\Delta - X^2\Pi$) is proportional to the growth rate of a-Si:H films so long as the gas pressure is held constant [39, 42]. It is assumed that this is closely related to the direct production of Si^* and SiH^* in the same way as SiH_3 and SiH_2 in the primary dissociation process of SiH_4 (see Fig. 2.10) [39]. On the other hand, no correlation has been recognized between the growth rate and the line intensity of H_2^* (6 021 Å: $3p^3\Pi$-$2s^2\Sigma$) and H^* (6 563 Å: 3^2D-2^2p^0, H_α), probably because these excited states

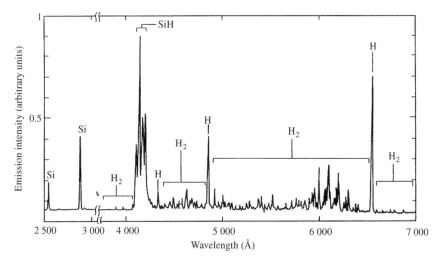

Fig. 2.13 Optical emission spectrum of a SiH_4 plasma [44, reproduced with permission]

require electron excitation as a secondary process after the dissociation of SiH$_4$. OES is a simple and information-rich diagnostic method, although it is effective only for light-emitting species.

(e) Ion mass spectrometry (IMS)

In this technique, a minute orifice (200 μm in diameter) is positioned at the center of the substrate mount (reactor anode), and species passing through this orifice are led in to and then analyzed by a mass spectrometer which is evacuated differentially [39]. It is essential to turn *Off* the ionizer within the mass spectrometer; this ensures that only the originally ionized species (cations) undergo exact analysis. If the ionizer is kept *On*, a large amount of SiH$_4$ passing through the orifice is dissociated before being ionized. As shown in Fig. 2.9, the threshold energy for SiH$_4$ dissociation by electron impact is far lower than that of SiH$_4$ ionization, i.e. it is by no means an easy task to carry out exact measurements on neutral species [49, 50]. The concentration of total cations:

$$\sum_i [X_i^+]$$

as measured with the ionizer *Off*, is nearly of the same order as the electron concentration in the plasma, or to be more exact, equal to the total concentration of electrons plus anions.

2.3 Growth Mechanisms

2.3.1 LIFETIMES AND DIFFUSION OF SPECIES IN THE SiH$_4$ PLASMA

Species generated within the SiH$_4$ plasma are involved in the formation of the topmost layer of an a-Si:H film through transport processes in the growing film (mainly diffusion) and surface processes on the film itself (such as sorption, surface diffusion and various reactions). As stated in Section 2.2, in the SiH$_4$ plasma under steady-state conditions, the concentration of the various species within the positive column is a few orders of magnitude lower than the concentration of the parent molecules, ([SiH$_4$] = 10^{15} cm^{-3}). For this reason, it may be claimed that the secondary processes involving species produced by the dissociation of SiH$_4$ following inelastic collision with electrons, mostly result from collision with SiH$_4$.

Table 2.3 shows the reactions that occur between SiH$_4$ and the major species found in an SiH$_4$ plasma, plus their reaction rate constants (k) and reaction lifetimes (τ) at gas pressures of 50 m torr [38, 39, 51, 52]. The most conspicuous feature is that SiH$_3$ is very stable after its collision with SiH$_4$, and its lifetime in the SiH$_4$ plasma is significantly longer than that of any of the other radicals and atoms. This is also partly responsible for the fact that the concentration of SiH$_3$ in the steady state is the highest among the various species, i.e. [SiH$_3$] = 10^{12} cm^{-3}. It should be noted that in estimating [SiH$_3$], equation (2.8) was substituted by

Table 2.3 Rate constants and reaction lifetimes for the various reactions between SiH$_4$ and other species in the SiH$_4$ plasma at a gas pressure of 50 mtorr [39]

Reaction	Rate constant k (cm^3 s^{-1})	Lifetime τ(s)
$\boxed{H}$ + SiH$_4$ $\longrightarrow$ H$_2$ + SiH$_3$	5×10^{-12}	1.1×10^{-4}
$\boxed{Si}$ + SiH$_4$ $\longrightarrow$ Si$_2$H$_4$(Si$_2$H$_4$ + SiH$_4$ $\longrightarrow$ Si$_2$H$_6$)	—	—
$\boxed{SiH}$ + SiH$_4$ $\longrightarrow$ Si$_2$H$_5$(Si$_2$H$_5$ + SiH$_4$ $\longrightarrow$ Si$_2$H$_6$)	3.3×10^{-12}	1.72×10^{-4}
$\boxed{SiH_2}$ + SiH$_4$ $\longrightarrow$ Si$_2$H$_6$	2.3×10^{-10}	2.47×10^{-6}
$\boxed{SiH_3}$ + SiH$_4$ $\longrightarrow$ SiH$_3$ + SiH$_4$	$\sim$0	Very long

the simpler equation (2.9). While in the third term of the right-hand side of equation (2.8):

$$\sum k_{ij} N_j N_i$$

the contribution of $N_j(= [\mathrm{SiH_4}])$ is generally not negligible, the rate constant k for SiH$_3$ is essentially zero, as shown in Table 2.3, i.e. $k_{ij} = 0$, and hence approximation by using equation (2.9) is reasonable. If this property of SiH$_3$ is utilized, it may become possible to selectively feed SiH$_3$ alone to the substrate by spatially separating the substrate electrode from the plasma column, as will be described later. When no electrons are present (with the plasma *Off*), the time change in [SiH$_3$] is determined by diffusion and collision reactions between the SiH$_3$ species. According to the detailed experimental data obtained by Itabashi *et al.* [53] using IRLAS, the diffusion coefficient, D, of SiH$_3$, under pressures of 50 m torr of SiH$_4$ and 30 m torr of H$_2$: can be calculated as follows:

$$D(\mathrm{SiH_3 \ in \ H_2/SiH_4}) \cong 2.5 \times 10^3 \ \mathrm{cm^2 \ s^{-1}} \tag{2.10}$$

2.3.2 ADSORPTION/DESORPTION OF SPECIES AND GROWTH RATES

Species, such as SiH$_3$, arriving at the surface of a growing a-Si:H film are partly reflected (reflection, 1-β), partly adhere through surface diffusion and reaction (sticking, s) and partly desorbed after recombination (recombination, γ). These three coefficients, β, s and γ, are related in the following way:

$$\beta = s + \gamma \tag{2.11}$$

where β is designated the loss probability [54], i.e. β represents the probability of the arriving species disappearing by being incorporated into the a-Si:H film or by changing into other species (see Fig. 2.14).

Matsuda *et al.* prepared a-Si:H films by selectively using SiH$_3$ as the main radical species in a triode-type reactor and determined β and s [54]. In this set-up, a SiH$_4$ plasma is formed between the cathode and the mesh electrode, and

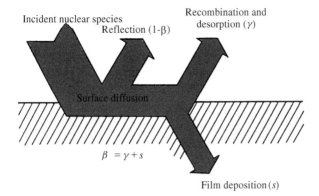

Fig. 2.14 A schematic diagram illustrating the surface processes occurring for the arriving species at the surface of a growing a-Si:H film

the as-generated radicals pass through the mesh to diffuse towards the substrate, with long-lived SiH_3 being left as the predominant radical species.

Two methods for measuring the loss probability β are illustrated in Fig. 2.15 [54]. The first of these is the classical grid method, in which the thickness of the film deposited on the substrate (e_1) and that of the film deposited on the underface (e_2) of a grid placed above the substrate e_1, and β are related by the following expression [55]:

$$e_2/e_1 = 1 - \beta \qquad (2.12)$$

The distance between the grid and the substrate is set shorter than the mean free path of the gas molecules. In the second method, a film is deposited on to a Si wafer substrate on which micron-sized trenches have been cut, and β is then estimated from the thickness profile of the trench cross-section (the step-coverage method) [54]. In this case, a film-thickness profile is obtained by a computer simulation with β as a parameter, and then compared with the experimental data obtained from scanning electron microscopy (SEM) cross-sectional images. In Fig. 2.16, the surface loss probability β of SiH_3 is plotted as a function of the substrate temperature T_S during the deposition of a-Si:H. The mean value of β for temperatures from room temperature to 480°C is found to be as follows [54]:

$$\beta = 0.26 \pm 0.05 \qquad (2.13)$$

Therefore, it can be seen that, more than 70% of the SiH_3 that arrives at the surface is reflected.

The sticking coefficient s can be determined from the dependence of the growth rate of the a-Si:H film (or rate of increase of the film thickness e_0 in Fig. 2.15) on the substrate temperature T_S (see Fig. 2.17). This figure shows changes in the bonded-hydrogen content, C_H, of the film (atom%) together with the film growth rate, as a function of the substrate temperature. As the hydrogen content of the film becomes almost zero at $T_S = 480°C$, it is assumed that the coverage factor

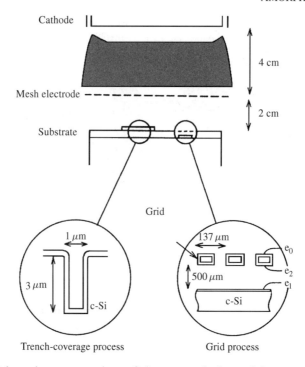

Fig. 2.15   Schematic representations of the two methods used for measuring the loss probability β (reflection: $1 - \beta$) with a triode-type reactor and selective SiH$_3$ supply: (a) trench-coverage method; (b) grid method [54, 55, reproduced with permission]

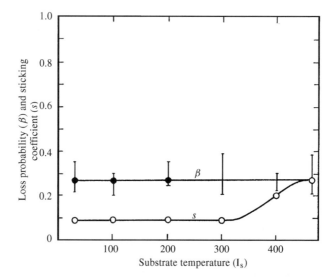

Fig. 2.16   Dependence of the loss probability $\beta(\bullet)$ and sticking coefficients $s(\circ)$ of SiH$_3$ on substrate temperature [54, reproduced with permission]

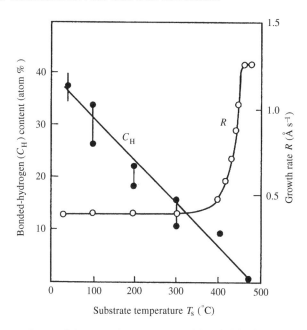

Fig. 2.17 Dependence of the growth rate R (o) and bonded-hydrogen C_H content (•) of an a-Si:H film on substrate temperature [54, reproduced with permission]

of the hydrogen on the surface of the growing film is considerably decreased at this temperature. In other words, it seems likely that the 'dangling bonds' of the Si are exposed at the surface of the growing film and capture the arriving and adsorbed SiH$_3$ with a very high probability. Under such circumstances, it can be assumed that the recombination coefficient γ is almost zero ($\gamma \sim 0$). The variation in the sticking coefficient s, as calculated from the data shown in Fig. 2.17, by assuming that $\gamma \sim 0$ at $T_S = 480°C$, as a function of temperature, is shown in Fig. 2.16 [54].

Let us now try to derive a quantitative estimation of the growth rate of a-Si:H from the concentration gradient of the SiH$_3$ radicals by using experimentally determined values of β and s. From the data obtained for the spatial distribution of [SiH$_3$] (shown in Fig. 2.12), it is possible to obtain the gradient in the vicinity of the anode, $(d[SiH_3]/dx)_{x=0}$. If a substrate is placed on the anode to deposit the a-Si:H, the following relationships hold for the growth rate R, the molecular flux density of SiH$_3$ on the growing surface $J(SiH_3)(x = 0)$, the surface-loss probability β of SiH$_3$, the sticking coefficient s, the density of a-Si:H ρ and the mass m of Si$_1$H$_{0.25}$:

$$\beta J(SiH_3) = D(SiH_3 \text{ in } SiH_4/H_2)\left\{\frac{d[SiH_3]}{dx}\right\}_{x=0} \tag{2.14}$$

$$R = J(SiH_3) \times \frac{sm}{\rho} \tag{2.15}$$

Putting equations (2.10) and (2.13), $d[\mathrm{SiH_3}]/dx = 5.6 \times 10^{11}$ cm^{-4}, $s = 0.1$, $m = 4.7 \times 10^{-23}$ g, $\rho = 2.2$ g cm^{-3} into equations (2.14) and (2.15), we obtain the following [43]:

$$J(\mathrm{SiH_3}) = 5.8 \times 10^{15} \text{ cm}^2 \text{ s}^{-1} \tag{2.16}$$

$$R(\text{a-Si:H}) = 1.1 \text{ Å s}^{-1} \tag{2.17}$$

This latter value correlates well with experimental measurements, i.e. 1.79 Å s^{-1}.

On the basis of the above discussion, it is reasonable to consider SiH$_3$ as being the major source species (precursor) of a-Si:H. Since the steady-state concentration of SiH$_2$ is far lower than that of SiH$_3$, it is difficult to regard SiH$_2$ as a predominant species in the deposition of a-Si:H. However, as it is supposed that the incorporation of SiH$_2$ into an a-Si:H surface is augmented by an insertion reaction (to be discussed later), under some conditions SiH$_2$ may be a competitor to SiH$_3$ [40].

2.3.3 SURFACE-PROCESS MODEL

The hydrogen content C_H of an a-Si:H film decreases as the substrate temperature T_S rises, falling to less than 1% at 480°C (see Fig. 2.17). On the other hand, the growth rate of a-Si:H is almost constant up to $T_\mathrm{S} = 300$°C, beyond which it rises steeply and becomes saturated at 480°C. In order to explain these results, more microscopic information is needed. A number of models, based on the chemistry of the materials, have been proposed, and one of these concerning the surface processes of the SiH$_3$ radical and its relationship to the hydrogen coverage of the growing surface will be described below.

This model assumes that the growing surface of an a-Si:H film is fully covered by hydrogen for T_s ranging from room temperature to 300°C; this has been confirmed experimentally [56]. The arriving SiH$_3$ radicals are captured by Si−H bonds (to be denoted as ≡SiH) on the surface through weak chemical adsorption with a probability β). For the microscopic model of this adsorption, a 3-center bond is considered to be one of the candidates [57]:

$$\begin{array}{ccccc} \equiv\mathrm{SiH} & + & \mathrm{SiH_3} & \longrightarrow & \equiv\mathrm{SiHSiH_3} \\ \text{(surface)} & & \text{(arriving radical)} & & \text{(weak chemical adsorption)} \end{array} \tag{2.18}$$

It is assumed that SiH$_3$ is able to undergo surface diffusion by migration to adjacent surface Si−H bonds through a thermal hopping mechanism. In this process, if a dangling bond (to be denoted as ≡Si−) occurs at an adjacent site, then SiH$_3$ will selectively jumps to it to form a Si−Si bond, thus forming cross-links with other surface Si species, with the release of H$_2$:

$$\begin{array}{ccccccc} \equiv\mathrm{Si-} & + & \mathrm{SiH_3} & \longrightarrow & \equiv\mathrm{SiSiH_3} & \longrightarrow & \equiv\mathrm{SiSiH} & + & \mathrm{H_2} \\ \text{(dangling bond)} & & \text{(surface diffusion)} & & \text{(bonding)} & & \text{(cross-linking)} & & \text{(H}_2 \text{ release)} \end{array}$$

$$\tag{2.19}$$

If a SiH$_3$ species meets another SiH$_3$ species before being 'caught' by a dangling bond, the two radicals will combine to form either a stable molecule of Si$_2$H$_6$, which will then be desorbed, or SiH_4 through the extraction of superficial hydrogen (this silane will also be desorbed)

$$\equiv SiH + SiH_3 \longrightarrow \equiv Si- + SiH_4 \qquad (2.20)$$

Therefore, in the lower temperature range (room temperature $< T_S < 300°C$), where almost 100% of the surface is covered by hydrogen, dangling bonds ($\equiv Si-$) are formed mainly through the hydrogen-extraction reaction shown in equation (2.20), followed by the formation of Si−Si bonds, shown by the left-hand side of equation (2.19) to complete the elementary process of film deposition. The left-hand side of equation (2.19) represents an exothermic reaction, while the reaction shown in equation (2.20) is an endothermic one. As these two processes occur in conjugation, the total thermal activation energy is likely to be very low [58, 59]. In this temperature range, therefore, the growth rate for film deposition does not have a significant dependence on the substrate temperature.

When T_S rises above 300°C, dangling bonds at the surface are mainly formed by the release of H_2 from two mutually adjacent Si−H bonds:

$$2 \equiv SiH \longrightarrow \qquad 2 \equiv Si- \qquad + \qquad H_2 \qquad (2.21)$$
$$\text{(surface)} \qquad \text{(dangling bond formation)} \qquad \text{(gas released)}$$

The growth rate is controlled by the rate of dangling-bond formation given in equation (2.21), and the recombination coefficient γ (for equation (2.20) and others) becomes close to zero as T_S rises. The thermal activation energy for the reaction shown in equation (2.21) has been calculated to be 2.7 eV through the analysis of the data shown in Fig. 2.17 [60].

If T_S is further increased to 460–480°C, the growth rate becomes saturated (see Fig. 2.17). Under this condition, hydrogen coverage of the surface seems to be considerably decreased, with the growth rate being controlled by the supply of SiH_3. When T_S rises above 480°C, deposition now starts to result from the thermal dissociation of SiH_4, moving from the realms of plasma CVD to those of thermal CVD.

2.3.4 SURFACE DIFFUSION AND NETWORK STRUCTURE

The mobility of the species adsorbed on the surface determines the properties of the network structure. If these species can diffuse over an adequate distance, the probability of finding energetically favourable site rises as a result of relaxation of the amorphous structure. The surface diffusion length ℓ of the radicals can influence the qualitative discussion;

$$\ell = \sqrt{2D_s \tau_s} \qquad (2.22)$$

$$D_s = v a_0^2 \exp\left(-\frac{E_s}{kT}\right) \qquad (2.23)$$

where D_s is the surface diffusion coefficient of the radicals, τ_s the surface residence time, v the frequency factor, a_0 the distance between sites and E_s the thermal activation energy of hopping. In order to achieve adequate structural relaxation, it is necessary to increase ℓ. However, as direct desorption of adsorbed SiH_3 and SiH_2 is not likely, τ_s is limited by the growth rate and is almost equal to the mean time required for forming a single atomic layer. In this way the problem is reduced to increasing D_s.

Since SiH_3 is only held on the surface by weak chemical adsorption (equation (2.18)), it can diffuse over the surface via site-to-site hopping; this is equivalent to a smaller E_s in equation (2.23). On the other hand, SiH_2 is believed to form strong $Si-Si$ bonds through the insertion reaction given by equation (2.24), thus effectively increasing E_s, and resulting in a D_s value which is far smaller than that of SiH_3.

$$\equiv SiH \quad + \quad SiH_2 \quad \longrightarrow \quad \equiv SiSiH_3 \qquad (2.24)$$
$$\text{(surface)} \quad \text{(arriving radical)}$$

This effectively means that SiH_3 is the preferred species for providing adequate structural relaxation and constructing a dense network. As it is necessary to ensure the validity of equation (2.18) as a condition for enabling the surface diffusion of SiH_3, the hydrogen coverage of the growing surface must be kept as high as possible.

On the basis of the above qualitative discussion, two conditions can be proposed for depositing films of good quality, namely (1) to feed selectively species (SiH_3), which are readily diffusable along the surface, on to the growing surface, and (2) to increase the hydrogen coverage of the growing surface. One way to meet condition (1) is use of the triode-type reactor shown in Fig. 2.15, which utilizes the difference in the reaction rate constants between SiH_2 and SiH_3 against SiH_4 (for selective supply) [39, 42]. Another approach is the use of photo CVD involving sensitization with mercury, which selectively generates SiH_3 [28]. On the other hand, condition (2) can be simply achieved by supplying a large amount of hydrogen atoms into the plasma, either by diluting SiH_4 with a large volume of H_2 (the hydrogen-dilution method) [39, 42], or by feeding hydrogen which has been excited in a separate plasma (the remote-plasma method) [11]. These methods have proved to be very effective in improving the optoelectronic properties of silicon–alloy materials, such as a-Si_xGe_y:H and a-Si_xC_y:H [39, 42].

An extreme case of structural relaxation is the transition from the amorphous to the crystalline state. In practice, if the degree of dilution with hydrogen is increased, a-Si:H is partially turned into the microcrystalline state (to be designated as μc-Si:H). In Fig. 2.18, the volume fraction X_C of microcrystallites in μc-Si:H, prepared under higher dilution conditions ($SiH_4/H_2 = 1/19$, 1/9 and 1/4), is plotted against the substrate temperature T_S used during the preparation [42]. When $T_S < 150°C$, D_S has a lower value in equation (2.23) and the amorphous structure prevails. As T_S rises, D_S increases owing to the temperature effect, and the (amorphous) structure becomes partly microcrystalline, changing into μc-Si:H. However, if T_S surpasses $400°C$, X_C

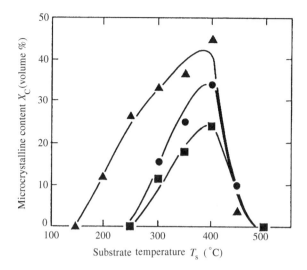

Fig. 2.18 Dependence of the volume fraction of microcrystalline silicon (X_C) in the film on substrate temperature, measured under three different plasma conditions [42, reproduced with permission]

begins to decrease again, with finally the whole volume becoming amorphous. This singular behavior is attributed to the decrease in the hydrogen coverage of the surface at higher temperatures, thus disabling site hopping under the adsorption conditions described by equation (2.18).

In this model, it is assumed that the key factors which promote structural relaxation are radicals supplied from the gas phase, substrate temperature and hydrogen-terminating dangling bonds on the surface. While the upper limit of the substrate temperature is defined by the requirements of hydrogen coverage, if hydrogen is replaced by another monovalent element, e.g. fluorine (F), it may be possible to realize adequate structural relaxation at higher substrate temperatures, on account of the stronger Si−F bond. However, there have been no reports which indicate that fluorinated amorphous silicon (designated as a-Si:F) is significantly better than a-Si:H, from the viewpoint of the low-temperature process and other conditions of preparation [61].

With regard to the role of hydrogen in the plasma for the formation of μc-Si:H, a model involving the etching of weak Si−Si bond during the growth process has been proposed [62], in addition that described above. Under certain conditions, it may be possible that the two mechanisms operate concurrently.

While the qualitative discussion so far has been based on a two-dimensional model which assumes the diffusion of radicals along the growing surface, another idea has been proposed in which a few atomic layers on the surface are regarded as three-dimensional growth zones through which hydrogen diffuses to relax the network structure [63]. In the latter model, hydrogen is released (or extracted), not only by the mechanisms described by equations (2.19)–(2.21), but also by release from the bulk material in the vicinity of the surface. These two concepts do not contradict each another, and moreover are expected to converge in the near future as a result of more precise studies of the surface processes.

References

1. *Applied Physics Handbook*, edited by the Japanese Society of Applied Physics, Maruzen, Tokyo (1990) Chapter 11, 587 (in Japanese).
2. J. C. Phillips, *J. Non-Cryst. Solids*, **34** (1979) 153.
3. G. H. Döhler, R. Dandoloff and A. Bilz, *J. Non-Cryst. Solids*, **42** (1980) 87.
4. Y. Kanemitsu and H. Kuroda, *Oyo Buturi (Jpn J. Appl. Phys.)*, 54 (1985) 1154 (in Japanese).
5. *Glass Handbook*, edited by S. Sakuhana, *et al.*, Asakura Publishers, Tokyo (1975) (in Japanese).
6. *Fundamentals of Amorphous Semiconductors*, edited by K. Tanaka, OHM Publishing Company, Tokyo (1982) 5 (in Japanese).
7. *Fundamentals of Amorphous Metals*, edited by T. Masumoto, OHM Publishing Company, Tokyo (1982) (in Japanese).
8. *Hydrogenated Amorphous Silicon, Part A: Semiconductors and Semimetals*, Vol. 21, edited by J. I. Pankove, Academic Press, NY, (1984).
9. S. Iizima, H. Okushi, A. Matsuda, Y. Yamasaki, K. Nakagawa, M. Matsumara and K. Tanaka, *J. Appl. Phys.*, **19** (1980) Suppl, 521.
10. S. Hine, *et al.*, *Proceedings of the 13th Seminar on Properties and Applications of Amorphous Materials*, (1986) 41 (in Japanese).
11. G. Lucovsky and D. V. Tsu, *J. Non-Cryst. Solids*, **97/98** (1987) 265.
12. K. Tanaka, Y. Yamasaki, K. Nakagawa, A. Matsuda, H. Okushi, M. Matsumura and S. Iizuma, *J. Non-Cryst. Solids*, **35/36** (1980) 475.
13. M. H. Brodsky, R. S. Title, K. Weiser and G. D. Petit, *Phys. Rev. B : Solid State*, **1** (1970) 2632.
14. A. J. Lewis, G. A. N. Connell, W. Paul, J. R. Pawlik and R. J. Temkin, *Tetrahedrally Bonded Amorphous Semiconductors*, edited by M. H. Brodsky and S. Kirkpatrick, American Institute of Physics, New York, (1974) 27.
15. T. D. Moustakas and W. Paul, *Phys. Rev. B: Solid State*, **15** (1977) 1564.
16. M. Pinarbasi, N. Marley, A. Myers and J. R. Abelson, *Thin Solid Films*, **171** (1989) 217.
17. M. Hirose, *Hydrogenated Amorphous Silicon, Part A: Semiconductors and Semimetals*, Vol. 21, edited by J. I. Pankove, Academic Press, NY, (1984) 109.
18. B. A. Scott, *Hydrogenated Amorphous Silicon, Part A: Semiconductors and Semimetals*, Vol. 21, edited by J. I. Pankove, Academic Press, NY, (1984) p. 123.
19. Y. Mishima, M. Hirose, Y. Osaka, K. Nagamine, Y. Ashida, N. Kiragawa and K. Isogaya, *Jpn J. Appl. Phys.*, **22** (1983) L46.
20. N. Sol, D. Kaplan, D. Dienmegard and D. Dubreuil, *J. Non-Cryst. Solids*, **35/36** (1980) 291.
21. T. Suzuki, *Jpn J. Appl. Phys.*, **19** (1980) Suppl., 91.
22. M. Hirose, *J. Phys. (Paris)*, **10** (1981) Suppl., C4-705.
23. M. Hanafusa, *DENGAKUSHI*, **28** (1988) 46 (in Japanese).
24. M. Konagai, *MRS Symp. Proc.*, **70** (1986) 257.
25. H. Koinuma, *DENSHIKOGYO GEPPOU*, **28** (1986) 46 (in Japanese).
26. R. E. Rocheleau, *MRS Symp. Proc.*, **70** (1986) 37.
27. U. Itoh, Y. Toyoshima, H. Ohuki, N. Washida and T. Ibuki, *J. Chem. Phys.*, **85** (1986) 4867

28. Y. Uchida, *Hydrogenated Amorphous Silicon, Part A: Semiconductors and Semimetals*, edited by J. I. Pankove, Academic Press, NY, **21** (1984) 41.
29. Y. Toyoshima, K. Kumata, U. Itoh, K. Arau, A. Matsuda, N. Washida and G. Inoue, *Appl. Phys. Lett.*, **46** (1985) 584.
30. I. Yamada, *Hydrogenated Amorphous Silicon, Part A: Semiconductors and Semimetals*, edited by J. I. Pankove, Academic Press, NY, **21** (1984) 83.
31. A. H. Mahan, *MRS Symp. Proc.*, **219** (1991) 673.
32. J. Hanna, *MRS Symp. Proc.*, **149** (1989) 11.
33. Battelle Columbus Laboratories, *Amorphous Silicon/Materials Contractor's Review (Meeting Abstracts by SERI)* (May 1979).
34. B. Chapman, *Glow Discharge Processes: Sputtering and Plasma Etching*, John Wiley and Sons, Chichester (1980).
35. A. Matsuda, T. Kaga, H. Tanaka and K. Tanaka, *Jpn J. Appl. Phys.*, **23** (1984) L567.
36. N. Hata, A. Matsuda, K. Tanaka, K. Kajiyama, N. Mori and K. Sajiki, *Jpn J. Appl. Phys.*, **22** (1983) L1.
37. P. Kocian, *J. Non-Cryst. Solids*, **35/36** (1980) 201.
38. J. P. M. Schmitt, *J. Non-Cryst. Solids*, **59/60** (1983) 649.
39. A. Matsuda, *Researches of ETL*, Electrotech. Lab., Tsukuba, No. 864, (1986) 22 (1986) (in Japanese).
40. N. Hata, *Researches of ETL*, Electrotech. Lab., Tsukuba, No. 901, (1989) 7 (in Japanese).
41. K. Tachibana, *Proceedings of 8th Symposium on ISIAT'84 (Kyoto, 1984)*, 319.
42. K. Tanaka and A. Matsuda, *Mater. Sci. Rep.*, **2** (1987) 139.
43. N. Itabashi, N. Nishiwaki, M. Magane, S. Naito, T. Goto, A. Matsuda, C. Yamada and E. Hirota, *Jpn J. Appl. Phys.*, **29**, (1990) L505.
44. A. Matsuda and K. Tanaka, *Thin Solid Films*, **12** (1982) 171.
45. N. Itabashi, K. Kato, N. Nishiwaki, T. Goto, C. Yamada and E. Hirota, *Jpn J. Appl. Phys.*, **27** (1988) L1565.
46. H. Lee, J. P. de Neufville and S. R. Ovshinsky, *J. Non-Cryst. Solids*, **59/60** (1983) 671.
47. A. Matsuda, K. Nakagawa, K. Tanaka, M. Matsumara, S. Yamasaki, H. Okoshi and S. Iizima, *J. Non-Cryst. Solids*, **35/36** (1980) 183.
48. R. W. Griffith, F. J. Kampas, P. E. Vanier and M. D. Hirsh, *J. Non-Cryst. Solids*, **35/36** (1980) 391.
49. G. Turban and Y. Cathrine, *Thin Solid Films*, **48** (1979) 57.
50. R. Robertson, D. Hills, H. Chatham and A. Gallagher, *Appl. Phys. Lett.*, **43** (1983) 544.
51. M. Browrey and J. H. Purnell, *Proc. R. Soc. (Lond.)*, **A321** (1971) 667.
52. E. A. Austin and F. W. Lampe, *J. Phys. Chem.*, **81** (1977) 1134.
53. N. Itabashi, K. Kato, N. Nishiwaki, T. Goto, C. Yamada and E. Hirota, *Jpn J. Appl. Phys.*, **28** (1989) L325.
54. A. Matsuda, K. Nomoto, Y. Yakeuchi, A. Suzuki. A. Yuuki and J. Perrin, *Surf. Sci.*, **227** (1990) 50.
55. J. Perrin, *Appl. Phys. Lett.*, **50** (1987) 433.
56. G. H. Lin, J. R. Doyle, M. He and A. Gallagher, *J. Appl. Phys.*, **64** (1988) 188.
57. R. Fisch and D. C. Licciardello, *Phys. Rev. Lett.*, **41** (1978) 889.
58. J. Perrin, Y. Takeda, N. Hirano, Takeuchi and A. Matsuda, *Surf. Sci.*, **210** (1989) 114.
59. R. Robertson and A. Gallagher, *J. Chem. Phys.*, **85** (1986) 3623.
60. K. Tanaka and A. Matsuda, *Thin Solid Films*, **163** (1988) 123.
61. M. Janai, R. Weil and B. Pratt, *J. Non-Cryst. Solids*, **59/60** (1983) 743.
62. A. P. Webb and S. Veprek, *Chem. Phys. Lett.*, **62** (1979) 173.
63. I. Shimizu, *J. Non-Cryst. Solids*, **114** (1989) 145.

3 STRUCTURAL PROPERTIES

This chapter describes the structure of amorphous silicon from the view point of various aspects, such as the macroscopic morphology, content and spatial distribution of hydrogen in thin films, density, structural order, defects, impurities and doping, in close relationship to the preparation and growth conditions, as previously described in Chapter 2. In particular, with regard to the structural order and defects, basic considerations will be provided in comparison with crystalline silicon. The hierarchy of structural order, experimental methods to determine the order and their limitations, definition of defects, proposed structural models for defects and electron spin resonance (ESR) spectroscopic analysis, and the concepts of thermal and chemical equilibrium for defining the defect density will be discussed.

3.1 Morphology (Macroscopic Aspects)

As is well known, hydrogen-free a-Si prepared by evaporation or sputtering contains a large amount of microvoids of size $10-100$ Å [1]. As argued in detail in the preceding section, if the surface of the growing film is devoid of hydrogen, then the arriving species such as Si or Si_n can have only a very small surface diffusion coefficient (D_S), which suppresses the structural relaxation and causes microvoids to develop in the network. For the same reason, if the substrate temperature is lower, then D_S is reduced in a-Si:H, according to equation (2.23) and microvoids are readily produced. It is also known that if films are grown with SiH_2, which has a larger sticking coefficient, and hence a smaller D_S, using this silane as the precursor species in place of SiH_3 (having a higher D_S), voids and columnar structures are formed in the a-Si:H, irrespective of the substrate temperature [2]. Taking into consideration the fact that a-Si:H is a non-equilibrium material, it is quite natural that a-Si:H assumes varied structures with a close dependence on the growing process.

Knights prepared a-Si:H film while systematically changing several parameters, such as the SiH_4/Ar ratio, RF (13.56 MHz) power and substrate temperature (T_S), and examined the resultaning films by using transmission electron microscopy (TEM) [2]. As a general tendency, vertical columns of width $100-300$ Å were recognized to be growing from the substrate, when films were

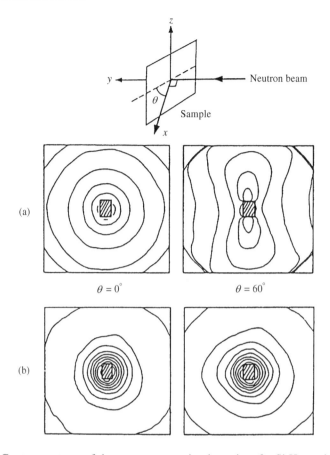

Fig. 3.1 Contour patterns of the neutron scattering intensity of a-Si:H, as viewed in the x–z plane: (a) $SiH_4/Ar = 3/97$, 25 W, and T_S = room temperature; (b) $SiH_4 = 100\%$, 2 W, and $T_S = 300°C$. The film prepared under the conditions (a) is anisotropic with respect to the incident angle of the beam [3, reproduced with permission]

deposited under conditions of smaller SiH_4/Ar ratios, higher RF power and lower T_S (i.e. close to room temperature). Fig. 3.1(a) shows the results of small-angle scattering analysis neutron beams and X-rays for the same film specimen prepared under these conditions [3]. According to the analysis, the marked anisotropy in scattering intensity suggests the presence of micro-rod structures, with diameters of ca 60 Å, arranged vertically with respect to the substrate [3]. On the basis of these facts, the following process may be inferred. First of all, a number of island structures with sizes of ca 100 Å are formed on the substrate in the initial process, and coalesce partly together. Then the, amorphous network begins to grow in the vertical direction, but owing to the lack of any transversal diffusion of the species, a form of so-called connective tissue linking these islands builds up vertically as a loose network, involving microvoids. It is conceived that hydrogen is locally concentrated among the connective tissue (loose network) joining the island structures (relatively dense network), as will be described later. Since such a film presents generally limited photoluminescence and photoconductivity, it is inappropriate for device applications.

On the other hand, a-Si:H film prepared under conditions of higher SiH_4/Ar ratios, lower RF power and higher T_S (i.e. 300°C) has no vertical rod structures that are discernible either by TEM or neutron scattering analysis (Fig. 3.1(b)). These results confirm that film preparation conditions which ensure a greater surface diffusion coefficient D_S of the species discussed in Chapter 2, provide adequate structural relaxation and a very small anisotropy of the film structure. It goes without saying that a-Si:H film grown under the latter conditions has higher emission efficiencies in photoluminescence, higher photoconductivities and higher electron mobilities [4].

The effectiveness of the triode reactor and the hydrogen-dilution method described in the preceding chapter has been demonstrated by observations of cross-sections of a-Si:H films using TEM: films prepared by using a diode reactor present vertical rod structures, while those grown in a triode reactor contains none of these structures [5].

3.2 Hydrogen in a-Si:H Films

The hydrogen contained in a-Si:H films may be studied by various means. Representative methods include infrared absorption [6], secondary-ion mass spectrometry (SIMS) [7], nuclear-reaction method [8], hydrogen thermal evolution [9], proton–electron nuclear double resonance (^{1}H-ENDOR) spectroscopy [10] and proton nuclear magnetic resonance (^{1}H-NMR) spectroscopy [11]. Each of these methods possess their own advantages and disadvantages. The total hydrogen contents can be determined by nuclear reaction, thermal release, nuclear magnetic resonance spectroscopy and SIMS. Infrared absorption allows the collection of information on bonded hydrogen, while NMR spectroscopy is effective in locating the spatial distribution of the hydrogen. From an experimentalist's point of view, the infrared absorption technique, which is readily applicable to film specimens formed on high-resistance Si crystals, seems to be simple and yet able to provide a good deal of information.

To summarize the experimental results, it may be concluded that a-Si:H film contains from a few to some 20 odd atomic (atom.)% of hydrogen in two different phases depending upon the method of preparation, with most of these forming Si—H bonds, while $\leqslant 1$ atom% exist as free hydrogen molecules [12]. The findings obtained by infrared absorption, NMR spectroscopy and thermal release will be described in detail below.

3.2.1 SILICON–HYDROGEN BOND TYPES

The infrared absorption technique provides detailed information about the type of bond between silicon and hydrogen. Table 3.1 shows various types of Si—H

Table 3.1 Wavenumbers for the different vibrational modes of various types of Si−H bond [6]

Structural group	Vibrational mode (cm^{-1})		
(type of Si−H bond)	Stretching	Bending	Wagging
SiH	2000		630
SiH$_2$	2090	880	630
(SiH$_2$)$_n$	2090−2100	890, 845	630
SiH$_3$	2140	905, 860	630

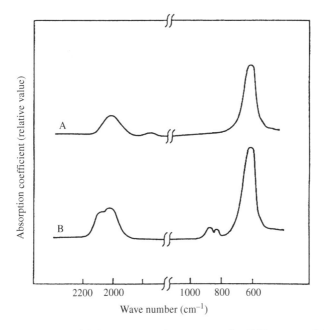

Fig. 3.2 Two examples of infrared absorption spectra of a-Si:H measured for different T_S values: (A) 250 °C; (B) room temperature [14, reproduced with permission]

bonds, their vibrational modes and the wavenumbers of the infrared absorptions which corresponds to them [6]. While the data given in Table 3.1 have been generally accepted, identification of the absorption wavenumber for SiH$_x$($x \neq 1$) may be somewhat arguable, in particular for the case of a-Si:H prepared by reactive sputtering [13]. Fig. 3.2 shows some typical infrared spectra of a-Si:H [14]. It is evident from the presence of the absorption bands at around 2100 cm^{-1} (stretching mode) and at around 800–900 cm^{-1} (bending mode) [14] that films grown at room temperature contain not only monohydride (Si−H), but also dihydride species such as Si=H$_2$, (Si=H$_2$)$_n$ and Si≡H$_3$. On the other hand, in films grown at $T_S = 250°C$, the Si−H type bond is predominant. In Fig. 3.3, the absorption coefficient α for various vibrational modes is plotted against T_S

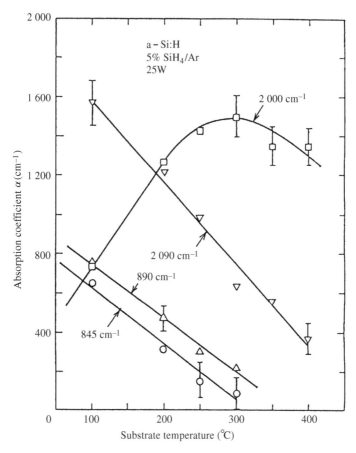

Fig. 3.3 The dependence of infrared absorption in various Si−H vibrational modes on substrate temperature [6, reproduced with permission]

[6]. Knights suggested, by using the two-dimensional structural model shown in Fig. 3.4, that if the a-Si:H network includes $Si=H_2$ or $(Si=H_2)_n$ chains, that voids and dangling bonds can be readily formed around them [2]. In other words, the hydrogen serves not always for reducing the amount of defects; moreover it matters more significantly how the hydrogen is incorporated (to be discussed later).

It is possible to determine the bonded-hydrogen content C_H from the areas of the various bands in the infrared absorption spectrum. For example, C_H can be obtained from the integrated absorption of the Si−H band around 630 cm^{-1} (wagging mode) by using the following expression [4]:

$$C_H = A \int [\alpha(\omega)/\omega]\, d\omega \qquad (3.1)$$

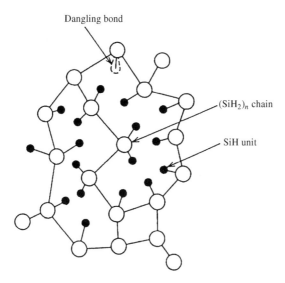

Fig. 3.4 A structural model of a-Si:H depicted as a two-dimensional network; broken lines represent dangling bonds [2, reproduced with permission]

where $\alpha(\omega)$ is the absorption coefficient at frequency ω, and the constant of proportionality A is inversely proportional to the absorption cross-section, and has been empirically determined [15] as:

$$A = 1.6 \times 10^{19} \text{ cm}^{-2} \tag{3.2}$$

For the evaluation of A, the following values have been widely adopted: 1.4×10^{20} cm^{-2} for the absorption band at around 2000–2100 cm^{-1} (stretching mode) [16], and 2.0×10^{19} cm^{-2} for the absorption band at 800–900 cm^{-1} (bending mode) [16]. The total content of bonded hydrogen calculated by using expressions (3.1) and (3.2) decreases as the substrate temperature is raised, and approaches zero at $T_S > 500°C$ (see Fig. 2.17 in Chapter 2).

The value of C_H obtained by this method closely agrees with the total content of bonded hydrogen as determined by the nuclear reaction method [8].

3.2.2 SPATIAL DISTRIBUTION OF HYDROGEN

With regard to the spatial distribution of hydrogen in a-Si:H, proton nuclear magnetic resonance (^{1}H-NMR) spectroscopy provides some useful information [11]; hydrogen has a nuclear spin I of 1/2 and thus gives high sensitivity in NMR, i.e. the gyromagnetic ratio γ is high. Reimer noted that the linewidth of free-induced attenuation for proton NMR spectroscopy of a-Si:H depends upon the dipole interactions among the hydrogen nuclei distributed in space, and suggested that the width of the NMR line directly provides information about the spatial distribution of the hydrogen [11].

If the linewidth is represented by the full-width-at-half-maximum (FWHM), the latter and the second moment M_2 are given, respectively [11], by:

$$\text{FWHM} = 2.36\sqrt{M_2} \tag{3.3}$$

$$M_2 = \frac{3}{5}\gamma^4\hbar^2 I(I+1)\sum_{ij} r_{ij}^{-6} \tag{3.4}$$

where I is the nuclear spin (equal to $1/2$ for a proton), γ is the gyromagnetic ratio, $\hbar = h/2\pi$ (h is the Planck constant) and r_{ij} is the distance between two nuclear spins i and j. It is evident from expressions (3.3) and (3.4) that the linewidth of the ^{1}H-NMR spectrum of a-Si:H is proportional to the local concentration of the protons in a three-dimensional network (and not to the mean concentration). If the hydrogen nuclei are spatially packed (concentrated hydrogen), the line becomes broader, and when protons are sparsely distributed (distributed hydrogen), the linewidth is reduced.

Figure 3.5 shows a typical ^{1}H-NMR spectrum (56.4 MHz) of a-Si:H [11], which characteristically contains both broad- and narrow-linewidth components.

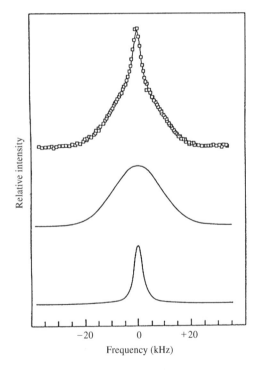

Fig. 3.5 A typical proton NMR spectrum of a-Si:H. The experimental plot (top) can be split into two components, i.e. a broad linewidth component of the Gaussian type (middle) and a narrow linewidth component of the Lorentzian type (bottom) [11, reproduced with permission]

This feature has been observed in other a-Si:H samples prepared at different substrate temperatures. The broad linewidth component is approximated by a Gaussian-type distribution curve with an FWHM of 20–30 kHz, with the narrow linewidth component being approximated a Lorentzian-type distribution curve of 2–4 kHz FWHM. It is possible to determine the hydrogen content from the area under the curve. Let us denote the hydrogen content (atom%) for the broad linewidth component by $[H]_b$, and that for the narrow linewidth component by $[H]_n$. It may be supposed that the former is attributed to concentrated hydrogen in the form of $(SiH_2)_n$ chains or SiH_x ($x = 1, 2, 3$), while the latter is attributed to hydrogen of the SiH type [11]. It should be noted, however, that hydrogen in the form of SiH is contributing to both the $[H]_b$ and $[H]_n$ components.

Figure 3.6 shows plots of $[H]_b$ and $[H]_n$ against the total hydrogen contents ($[H]_b + [H]_n$) for a number of a-Si:H samples prepared under various conditions. This figure includes data obtained for samples prepared by the glow discharge method, and also those produced by the reactive sputtering method. When the total hydrogen content exceeds 10 atom%, extra hydrogen contributes to $[H]_b$ (concentrated hydrogen) only, while $[H]_n$ (distributed hydrogen) is held at 6 atom% or less.

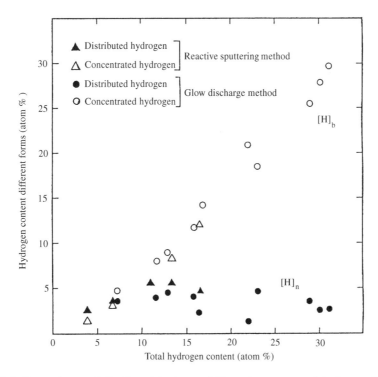

Fig. 3.6 Dependence of distributed hydrogen $[H]_n$ and concentrated hydrogen $[H]_b$ on the total hydrogen content, for a-Si:H films prepared by different methods and under various conditions [2, reproduced with permission]

These facts suggest that even in a-Si:H films prepared at high substrate temperatures, where no anisotropic structures are detected by TEM or neutron scattering investigations, the network includes two different phases with respect to the spatial distribution of hydrogen.

The total hydrogen content, $[H]_b + [H]_n$, as determined by the ^{1}H-NMR spectroscopy includes virtually all of the hydrogen whether or not it forms a Si$-$H bond. It is related, therefore, to the bonded-hydrogen content C_H, as determined by infrared absorption measurements, in the following way:

$$C_H \leqslant [H]_n + [H]_b \tag{3.5}$$

However, as the content of hydrogen in its molecular form is $\leqslant 1$ atom% (to be considered later) [12], it may be assumed that $C_H \approx [H]_b + [H]_n$. For the dependence of $[H]_b$ and $[H]_n$ on the substrate temperature T_S, see Fig. 3.19 in Section 3.5.3.

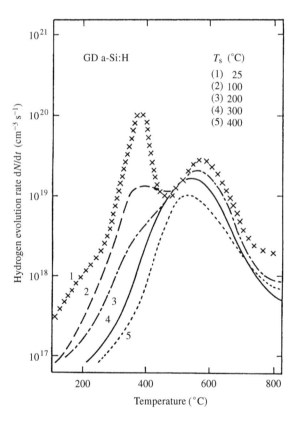

Fig. 3.7 Hydrogen evolution rates of a-Si:H samples prepared by the glow discharge method for different substrate temperatures; the heating rate is fixed at 20 °C min^{-1}. [18, reproduced with permission]

3.2.3 THERMAL STABILITY OF HYDROGEN

When a-Si:H is heated in vacuum at a fixed rate, e.g. $20°C$ min^{-1}, hydrogen begins to be released from the film above a certain temperature. The hydrogen evolution rates at various temperatures are plotted against the temperature (as a hydrogen evolution curve) in Fig. 3.7 [18]. While samples of a-Si:H prepared at lower T_S values display two peaks, in the cases where T_S is higher than $200°C$, the peak on the lower-temperature side is eliminated to produce a single-peaked curve. Fig. 3.8 shows the dependence of the hydrogen evolution curve peak temperature on the sample thickness (d) [18]. While the lower-temperature-side (LT) peak occurs at a fixed temperature irrespective of the sample thickness, that on the higher-temperature side (HT), denoted by T_M, rises as a function of the sample thickness d, indicating that the evolution rate is controlled by the diffusion of hydrogen atoms [18]. Experiments carried out with stacked films by using hydrogen and deuterium suggest the major diffusing species for the LT peak to be thermally formed hydrogen molecules.

On the basis of measurements of the HT peak temperature T_M for a-Si:H samples of different thickness d, prepared at $T_S = 25$, and $300°C$, Beyer and

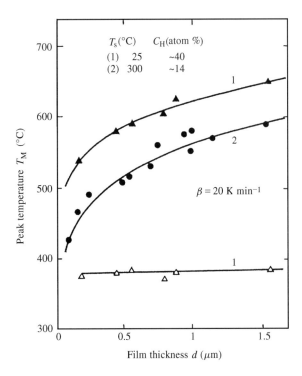

Fig. 3.8 The dependence of the hydrogen evolution curve peak temperature on film thickness for samples prepared at $T_S = 25 °C$ and $300 °C$: (▲, ●) low-temperature side; (△) high-temperature side [18, reproduced with permission]

Wagner derived the hydrogen diffusion coefficient D and diffusion energy E_D of (atomic) hydrogen [19], and determined D_0 and E_D by plotting $\ln(d/T_M)^2$ against $(1/T_M)$ [12, 18, 19]:

$$\ln(D/E_D) = \ln(\beta/k)(d/\pi T_M)^2 \tag{3.6}$$

$$D = D_0 \exp(-E_D/kT) \tag{3.7}$$

In equation (3.6), β represents the heating rate, and k is the Boltzmann constant. It has been shown that $D_0 = 100$ cm^2 s^{-1} and $E_D = 2.3$ eV for a-Si:H samples prepared at $T_S = 25°C$, and $D_0 = 3 \times 10^{-2}$ cm^2 s^{-1} and $E_D = 1.6$ eV for samples prepared at $T_S = 300°C$.

The diffusion of hydrogen with respect to the HT peak can be regarded as a combined process with simultaneously occurring hydrogen release from Si—H bonds and rearrangement of adjacent Si—Si bonds. Accordingly, the thermal release of a large amount of hydrogen does not necessarily leave a lot of dangling bonds in the film (see Section 3.5). The formation and diffusion of hydrogen molecules on the LT peak-side constitute an exothermic process, despite re-arrangement of the Si—Si bonds, as the formation of H_2 releases 4.5 eV bonding energy. Since the LT peak is observed only in samples prepared at lower substrate temperatures (as shown in Fig. 3.7) it is speculated that H_2 molecules are formed from concentrated hydrogen $[H]_b$ located in the voids and these readily diffuse among the columnar structures perpendicular to the film surface (see Section 3.5). The mean distance between the hydrogen atoms in the concentrated state is less than 1.9 Å [11]. In this case, too, it is known that the density of the film increases, or the network becomes denser, owing to the structural relaxation, as adequate heat is released through the formation of H_2 before the tempera-ture closely approaches the HT peak value [18]. This is one of the reasons why hydrogen molecules are not the major diffusing species in the higher-temperature region where the HT peak occurs [18].

3.3 Density

3.3.1 HYDROGEN-FREE AMORPHOUS SILICON

At about the beginning of the 1970s, there were many arguments, both experimental and theoretical, regarding the density of hydrogen-free a-Si prepared by the evaporation and sputtering techniques [1]. Among these, the ideas based on Polk's continuous random network (CRN) model were the most widely discussed [20].

This model, which will be described in detail in the next section (3.4), is a typical one for ideal amorphous structures, and is characterized by the complete absence of any dangling bonds in the network. In a CRN model introduced by Polk which consisted of 440 atoms, the highest density of a-Si (ρ_a) is $0.97\rho_c$, where ρ_c is the density of crystalline silicon (2.33 g cm^{-3}) [20]. Subsequently,

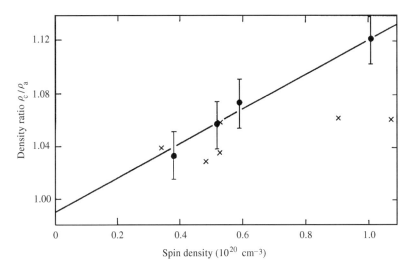

Fig. 3.9 The ratio of the density of c-Si to that of a-Si plotted against the spin concentration in a-Si. The hydrogen-free sample has been prepared by the electron beam evaporation method; film thickness is 2000 (●) or 8000 Å (×) [21, reproduced with permission. Copyright 1972 by the American Institute of Physics]

Polk constructed another CRN model including 519 atoms, using computer techniques, and obtained the following value:

$$\rho_a = 0.99\rho_c \qquad (3.8)$$

On the other hand, Brodsky and coworkers reported data for the systematic measurement of ρ_a and the ESR spin density, N_S of a-Si prepared by the electron beam evaporation method under various conditions [21]. These data show that a-Si samples prepared at higher substrate temperatures and lower growth rates have lower values of N_S, i.e. lower concentrations of dangling bond, with ρ_a as high as 0.97 ρ_c. In this discussion, the absence of dangling bonds means the elimination of voids, since the presence of a void within a-Si corresponds (in a one-to-one manner) to the formation of a dangling bond within the material. These data are shown in Fig. 3.9, where the extrapolation of the line indicates an ideal a-Si structure with dangling bonds completely eliminated, having $\rho_a = 1.01\rho_c$ [21].

While the extrapolation value for Brodsky's data does not coincide with Polk's model, which is based on a computer simulation with the absence of distortion, a-Si samples with density values which are very close to that of crystalline silicon (c-Si) have been obtained experimentally and studied theoretically.

3.3.2 HYDROGENATED AMORPHOUS SILICON

The density of a-Si varies markedly, if it contains hydrogen, depending upon both the content and the state of the hydrogen. The densities of different samples of a-Si:H have been measured by using a variety of techniques, including ^{15}N resonant nuclear reaction [22], flotation methods using solutions of known density [9, 24], and α-ray elastic scattering [25]. These results are summarized in Fig. 3.10, in the form of plots of density ratio against hydrogen content.

As the hydrogen content increases, ρ_a decreases but with the values showing a considerable variation. This indicates that the density depends on a number of factors, such as the state of the hydrogen within the films, the presence of voids of varying size, and strain in the films. Samples of a-Si:H with densities higher than 90% of that of crystalline silicon ($\rho_a > 0.9\rho_c$) are obtained only when prepared at substrate temperatures higher than 200°C. The highest value ever achieved is $\rho_a = 0.97-0.98\rho_c$. In Fig. 3.10, data from Weaire and coworkers represent the calculated values for a CRN model of an a-Si:H structure consisting of 397 silicon

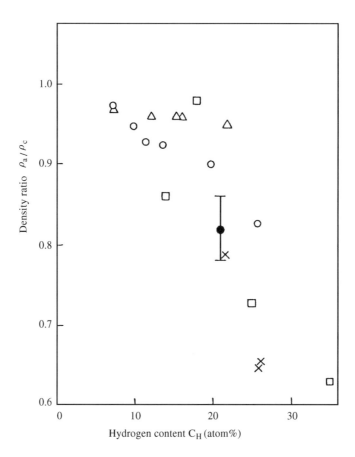

Fig. 3.10 Density ratio of a-Si:H to c-Si plotted against the hydrogen content of a film sample: (○) Fritzsche *et al.* [23]; (△): Tsuji and Minomura [24]; (□) Brodsky *et al.* [22]; (×) John *et al.* [25]; (●) Weaire *et al.* [26] [22–26, reproduced with permission]

and 83 hydrogen atoms, constructed by hand via a method similar to that used by Polk [26]. This model contains no dangling bonds within the structure except for those on the surface. Weaire and coworkers point out that when a hydrogen atom is added to the structure, dangling bonds are readily formed around it, and thus attempting to make a CRN model without dangling bonds gives a network structure containing multiple hydrogen atoms which are closely packed [26].

3.4 Structural Order

3.4.1 AMORPHOUS STRUCTURE AND HIERARCHY OF ORDER

So far, the structure of a-Si:H has been discussed in terms of a relatively coarse resolution over a few tens of angstroms. This section concerns the microscopic structure on the atomic scale, which is directly related to the fundamental electronic properties of the amorphous solid. As stated several times in the previous discussions, the amorphous structure is in a thermodynamically non-equilibrium state, and in principle can assume infinitely different states depending upon the preparative process. Microscopically speaking, the atomic arrangement is this case is not unique as in crystalline materials, and an enormous amount of structural disorder can take place. As the long-range order (LRO) of the atomic arrangement (or network structure as opposed to atomic array) is completely lost, a discussion based on reciprocal lattice space is meaningless. The only way to describe the amorphous structure is to treat statistically the structural order (or nature of disturbance) in real space.

The concept of statistical distribution is one way of describing the static structure of disordered systems. In this case, one of the most important variables is a geometrical parameter which allows us to discuss the disposition of atoms as a function of their spatial distance from a specified point. In fact, as described in Chapter 1, the structural order of amorphous solids has a definite hierarchy, where the order is disturbed increasingly as follows: SRO (short-range order) $\longrightarrow$ IRO (intermediate-range order) $\longrightarrow$ LRO (long-range order). A parameter is thus needed to specifically describe this situation.

Figure 3.11 shows a three-dimensional model which illustrates the geometrical parameters which play key role in describing the structural order of tetrahedral amorphous semiconductors. This model contains six geometrical parameters: the interatomic distance for the nearest neighbors (bond length), r_1, the bond angle θ ($\theta = 109.28°$ for sp^3 hybrid orbitals), the interatomic distance for second-nearest neighbors, r_2, the dihedral angle ϕ, the interatomic distance for third-nearest neighbors, r_3, and a ring-structure 'parameter' (for a five-membered ring). The statistical distributions of r_1 and θ (consequently, r_2 and the coordination numbers of the nearest neighbors are also included) are used for indexing the SRO, while the statistical distributions of ϕ, r_3 and the ring structure (ring statistics for five-, six-, and seven-membered rings) are used to index the IRO [27, 28]. There have been a number of attempts to build up amorphous structures theoretically by using these parameters. It is not easy, however, to take the process in the reverse order, i.e. to

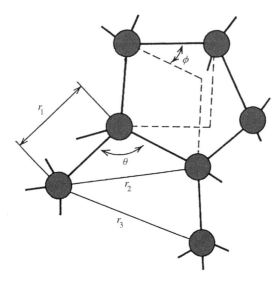

Fig. 3.11   Three-dimensional model illustrating the parameters used for describing the structural order of tetrahedral amorphous semiconductors

determine the statistical distribution of the various parameters from experimental data. The most useful techniques include X-ray diffraction and neutron diffraction, extended X-ray absorption fine structure (EXAFS) and Raman scattering. Some experimental results obtained in these ways and their interpretations in the case of hydrogen-free a-Si and a-Si:H are described below.

3.4.2 HYDROGEN-FREE AMORPHOUS SILICON

The presence of short-range order in amorphous solids can be verified by diffraction studies. When the scattering intensity of X-rays is measured as a function of the scattering angle γ, the structure factor $S(q)$ is obtained as a function of the wave vector $q = (4\pi \sin \gamma)/\lambda$. Fourier transform of the structure factor results in a pair-distribution function $g(r)$ for a two-body correlation between atoms or a radial distribution function $J(r)$, which is proportional to $r^2 g(r)$. Fig. 3.12 shows the radial distribution function (RDF), $J(r)$, of a-Si and c-Si obtained by the Fourier transformation of X-ray diffraction data [29].

The first peak corresponds to the Si–Si covalent bond length r_1 in Fig. 3.11. The value of r_1 for a-Si is nearly equal to the value of r_1 for c-Si ($r_1 = 2.34$ Å) within a 2–3% fluctuation range. The peak width is essentially determined by the thermal vibrations of the atoms and the quality of instrumentation being used. The number of Si atoms located on a sphere of radius r_1, i.e. the coordination number of the nearest neighbours, is determined on the basis of the integration value of the first peak. The coordination number is four for both a-Si and c-Si. This is direct evidence for the presence of short-range order in a-Si.

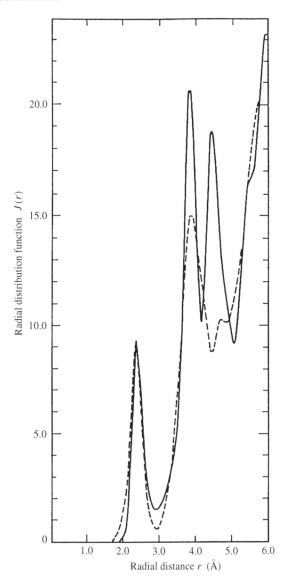

Fig. 3.12 Comparison of radial distribution functions for a-Si and c-Si: (- - - - -) a-Si; (————) c-Si prepared by crystallization of a-Si [29, reproduced with permission]

The second peak corresponds to the interatomic distance for the second-nearest neighbour, r_2. The peak area for c-Si suggests that a sphere of radius r_2 has 12 Si atoms, which is similar to the crystalline structure of diamond. In a-Si, however, the intensity of the second peak corresponding to r_2 is reduced and the peak width becomes broader. While r_1 of a-Si is close to that of c-Si, with only a small fluctuation, for r_2 the fluctuation increases significantly, thus showing the loss of order.

As shown in Fig. 3.11, r_1, r_2 and θ are related by the following expression:

$$r_2 \cong 2r_1 \sin(\theta/2) \tag{3.9}$$

with the variance in r_2 of a-Si being attributed to that of the bond angle θ. While the third peak is often overlaid on the second, the two peaks can be separated by using a simple model to determine the distribution of θ, i.e. $g(\theta)$ [20, 28]. According to analysis, the variance can as much as $10°(|\Delta\theta|/\theta < 10\%)$.

The most obvious difference in $J(r)$ between a-Si and c-Si is that the third peak corresponding to r_3 in c-Si does not exist in a-Si. Since r_3 involves all three of the parameters, r_1, θ and the dihedral angle ϕ, it is generally difficult to isolate the distribution of ϕ. If the distance is increased still further, the peak intensity is reduced and the peak width is expanded, thus making interpretation impossible. In view of the X-ray diffraction studies, it is evident that long-range order is absent in a-Si. However, it is well known that the qualitative shape of $J(r)$ of a-Si (or a-Ge) in Fig. 3.12 is not affected by the preparative conditions for up to $r \sim r_3$. It is certainly possible, therefore, that r_3 and ϕ have some definite features in the a-Si network which are related to intermediate-range order.

Some structural models have been proposed to reproduce the features of $J(r)$ shown in Fig. 3.12 [20, 30]. A model which explains the absence of the third peak ($r = r_3$), using two structural units, is well-established: this staggered configuration ($\phi = 60°$) and an eclipsed configuration ($\phi = 0°$) [31]. (see Fig. 3.13). If a network is constructed by using a staggered configuration ($\phi = 60°$) alone, a diamond-like structure (face-centered cubic lattice) of c-Si is obtained. If staggered configurations are placed in three directions and an eclipsed configuration is placed in one direction around a single Si atom, a wurzite-like structure (hexagonal lattice) is obtained. In c-Si, r_3 represents the normal interatomic distance for a staggered configuration. Accordingly, the microcrystalline model, in which a-Si is regarded as an assembly of microcrystals of Si, cannot explain the absence of the third peak [27]. The eclipsed configuration seems to be more promising. If all of the Si−Si bonds are configured in the eclipsed form and a slight distortion ($1°28'$) in the bond angle θ is allowed, then a structure composed of five-membered rings, an "amorphon", is formed, as shown in Fig. 1.11, (Chapter 1). It should be noted that while it is impossible to fill the complete space with amorphons, the structural unit of the five-membered ring is made up only of eclipsed configurations (with $\phi = 0°$). For this reason, the lack of the peak corresponding to r_3 in $J(r)$ for a-Si is often discussed in terms of "ring statistics", i.e. how five-, six- and seven-membered rings are distributed in the network (crystalline silicon contains only 6-membered rings) [32]. Anyway, there is no essential difference between the direct indexing of the intermediate-range order (IRO), such as the dihedral angle ϕ and the third-nearest-neighbour distance r_3, and the more intuitional argument in terms of the geometrical structural units, such as the eclipsed configuration and ring statistics.

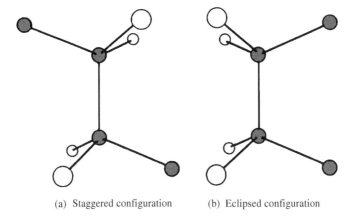

(a) Staggered configuration   (b) Eclipsed configuration

Fig. 3.13 Staggered configuration (a) with $\phi = 60°$, and eclipsed configuration (b) with $\phi = 0°$

Polk's model described in the preceding section (3.3.1) is a continuous random network (CRN) model, composed of 440 Si atoms and constructed by taking into account the following conditions: (1) no dangling bonds are left except for those on the surface, (2) $\Delta r_1/r_1$ is within $\pm 1\%$, (3) $\Delta\theta/\theta$ is within $\pm 10\%$, and (4) the distortion is held at a minimum level [20]. The model is characterized by the presence of 32 five-membered rings, in addition to 122 six-membered rings. The calculated value of the RDF ($J(r)$) is in good agreement with the experimental data shown in Fig. 3.12, and the disappearance of the peak at $r = r_3$ is successfully reproduced [20]. Later, Polk and Bondreaux attempted to modify this model and constructed another CRN consisting of 519 Si atoms by computer simulation. This model includes staggered and eclipsed configurations at a ratio of 2:1 in the network [33].

3.4.3 HYDROGENATED AMORPHOUS SILICON

Hydrogenated amorphous silicon, a-Si:H, which contains hydrogen in its network, is to be regarded as a substantially binary material, but both the SRO (r_1, θ, r_2) and IRO (ϕ, r_3) in the Si–Si network closely resembles those of hydrogen-free a-Si described in the preceding section.

From both X-ray diffraction [34] and neutron diffraction [35] data, the pair-distribution function $g(r)$, or the radial-distribution function $J(r)$ of a-Si:H, as obtained by Fourier transformations, qualitatively agree with the data obtained for a-Si shown in Fig. 3.12, in particular, with regard to the following aspects: (1) r_1 is nearly equal to the value found for of c-Si, (2) the FWHM of the r_2 peak is broadened, and (3) the third peak corresponding to r_3 of c-Si is missing. The most conspicuous difference is the fact that the apparent coordination number (the number of nearest-neighbour Si atoms around a Si atom) decreases as the bonded-hydrogen content C_H increases, owing to the presence of Si–H bonds. This is due

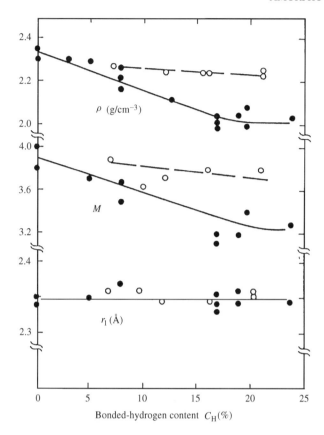

Fig. 3.14 Density (ρ), apparent coordination number (M) and nearest-neighbour spacing (r_1) as a function of the bonded-hydrogen content (C_H) of a-Si:H, as determined by X-ray diffraction measurements: (○) samples prepared by glow discharge; (●) sample prepared by reactive sputtering [24, reproduced with permission]

to the significantly smaller atomic scattering factor of hydrogen against X-rays, as a Si atom has 28 electrons while a H atom has only 1. Fig. 3.14 shows r_1, the apparent coordination number M, and the density ρ, measured for a-Si:H prepared by both the sputtering and glow discharge methods, as functions of C_H [24]. While the density and the apparent coordination number slowly decrease with an increase in C_H (see also Fig. 3.10 for the density data), r_1 remains at an almost constant value (2.35 ± 0.02), independent of C_H. The bond angle θ is nearly the same as that of a-Si, if determined by X-ray diffraction, but the Raman scattering, however, reveals a remarkable difference (to be discussed below). On the other hand, as hydrogen has a large atomic scattering factor towards neutron beams, neutron scattering provides some further information about the Si−H bond [35]. The length of the Si−H bond is reported to be 1.48 Å, while the distance between two hydrogen atoms in a =SiH$_2$ structural unit is 2.2 Å.

Useful information concerning the dihedral angle ϕ is difficult to obtain, as in the case of a-Si. According to the ring statistics of the 'hand-made' CRN model (consisting of 397 Si atoms and 83 H atoms) reported by Weaire et al. [26], discussed in Section 3.3.2, the number of ring structures per atom is 0.28, 0.41 and 0.83 for five-, six- and seven-membered rings, respectively [26]. While it is not necessary to regard these values as having particularly marked physical significance, it may be assumed that the order of the dihedral angle ϕ is more relaxed in a-Si:H, because the substantial constraints of the silicon network are reduced by the introduction of Si−H bonds.

As described above, the bond angle θ of Si, as determined by X-ray diffraction, is nearly identical for a-Si and a-Si:H. However, $\Delta\theta$ varies slightly depending upon the preparitive conditions, as clearly demonstrated by Raman scattering studies [28, 36, 37].

In both a-Ge and a-Si, the conservation of momentum for the lattice vibration is relaxed by the structural randomness, with all of the lattice vibrational modes becoming Raman-active [38]. Therefore, the first-order Raman scattering spectrum represents the spectrum for a lattice-vibrational density of states (phonons in crystals), weighted by transition matrix elements. The Raman scattering spectra of a-Ge and a-Si qualitatively resemble those of c-Ge and c-Si, respectively [28], indicating that the short-range order significantly affects the vibrational properties of the amorphous network.

Figure 3.15 shows first-order Raman scattering spectra of a variety of samples of a-Si and a-Si:H [36]. These may be qualitatively interpreted as blurred versions of the vibrational density of states for c-Si. A typical spectrum includes four energy bands: a transversal optical (TO) vibrational band at 475 ± 5 cm^{-1}, a longitudinal optical (LO) vibrational band at 380 ± 10 cm^{-1}, a longitudinal acoustic (LA) vibrational band at 310 ± 5 cm^{-1}, and a transversal acoustic (TA) vibrational band at 150 ± 5 cm^{-1}. The FWHM of the TO band is the broadest for hydrogen-free a-Si, i.e. 97 cm^{-1}, which is reduced to 80 cm^{-1} as the content of bonded hydrogen (C_H) increases. At the same time, the relative intensities of the TA, LA and LO bands diminish. These changes in the Raman spectra have been interpreted in the systematic work by Lannin and coworkers [28] and the model calculations by Beeman et al. [37]. According to these studies, the FWHM of the TO band, Δ_{TO}, is essentially proportional to the distribution width, $\Delta\theta$, of the bond angle θ of Si:

$$\Delta_{TO} \propto \Delta\theta \tag{3.10}$$

and in addition, Δ_{TO} is proportional to the ratio of the peak intensity of the TA band, I_{TA}, to the peak intensity of the TO band, I_{TO}:

$$\Delta_{TO} \propto I_{TA}/I_{TO} \tag{3.11}$$

The data presented in Fig. 3.15 show that when hydrogen is introduced into the a-Si network, Δ_{TO} decreases, or in other words, the bond angle fluctuation ($\Delta\theta$) is reduced, thus improving the SRO. As the coordination number for hydrogen

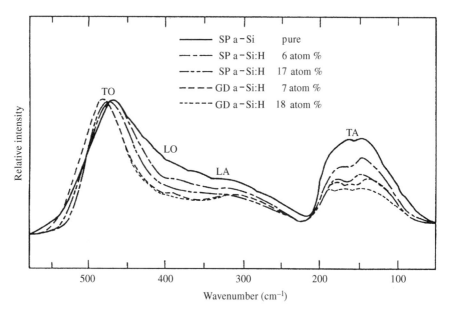

Fig. 3.15 Raman scattering spectra of various samples of a-Si and a-Si:H: (GD) samples prepared by glow discharge; (SP) sample prepared by reactive sputtering [36, reproduced by permission]

is unity, it may naturally be regarded that the distortion in the network is relaxed and the SRO is improved. At the present time, no other experimental technique is available that is as sensitive to $\Delta\theta$ as Raman scattering.

For the degree of order which covers a more extensive spatial area (intermediate-range order, IRO), there are two experimental techniques available which can provide useful information, namely small-angle X-ray scattering (SAXS) and small-angle neutron scattering (SANS). In X-ray diffraction, the scattering angle provides information on the degree of order at magnitudes of X-ray wavelengths, such as the interatomic distance of the nearest neighbours r_1, the bond angle θ, and the interatomic distance of the second-nearest neighbour r_2, if the CuK$_\alpha$ line is used. With very small incident and scattering angles, it is possible to obtain information concerning the fluctuations in electron density of the order of a few to several tens of interatomic distances [39].

Let us assume a simple two-phase system containing particles (having a different scattering length from a matrix) dispersed in the matrix at dilute concentrations. This system has interparticle spacings which are significantly larger than the size of a single particle, (r_p). Guinier demonstrated, by using this diluted two-phase model, that the scattering intensity $I(q)$ can be represented by the following expression, if $q r_p \ll 1$, where q is the wave vector:

$$I(q) \cong \exp(-q^2 R_g^2/3) \tag{3.12}$$

with $\quad |\boldsymbol{q}| = (4\pi/\lambda)\sin(\varepsilon/2) \cong 2\pi\varepsilon/\lambda$ (3.13)

and $\quad R_g^2 = \dfrac{1}{V_p}\displaystyle\int r^2 d^3r$ (3.14)

where λ is the wavelength of the X-ray or neutron beam, ε the small scattering angle, R_g the gyration radius, and V_g the volume of the particle. R_g is sometimes called the Guinier radius. If the particle is a sphere of radius R_S, then $R_g^2 = (3/5)R_S^2$, and hence:

$$I(\boldsymbol{q}) = \exp(-q^2 R_S^2/5) = \exp\left(-\frac{4\pi^2}{5\lambda^2}R_S^2\varepsilon^2\right)$$ (3.15)

If the logarithm of the scattering intensity $I(\boldsymbol{q})$ is plotted against the square of the small scattering angle (ε^2), it is possible to determine the radius of the particle from the gradient of the line. The expression represented in equation (3.12) is called Guinier's rule, with the plot of log $I(\boldsymbol{q})$ against q^2 being known as a Guinier plot.

In an actual system, density fluctuation and the distribution in shape and size of particles modeled for voids make it difficult to guarantee the linearity of a Guinier plot. In fact, $I(\boldsymbol{q})$ of a-Si:H (or a-SiGe:H) is very sensitive to the process and conditions of preparation [40]. In most cases, a Guinier plot takes the shape of a downward decreasing convex curve, with the minimum Guinier radius R_g of a typical a-Si:H sample being reported to be *ca* 10 Å [40].

While small-angle X-ray scattering (SAXS) is currently not a suitable method for examining the topology of the atomic arrangements, because of difficulty with data analysis, it is a very attractive method because it is very sensitive to the conditions of a-Si:H preparation, in contrast to the X-ray diffraction and Raman scattering techniques. The future deployment of SAXS in this field seems to be highly promising, as it can provide much important information on the intermediate-range order (IRO) in these amorphous materials.

3.5 Defects

3.5.1 CONCEPTS AND CLASSIFICATION OF DEFECTS

A crystal has a regular periodicity (translational symmetry) with respect to the atomic configuration, and a defect is defined as the displacement of an atom from its proper position in a perfectly regular lattice. The absence of an atom from its regular lattice point is called a vacancy, while an atom which is present outside the regular lattice is called an interstitial atom. If such definitions are applied to an amorphous solid, all of the atoms represent defects [41]. Since the amorphous network lacks long-range order, the relative positioning of the atoms is meaningful only in terms of a statistical distribution, (as described in Section 3.4.1).

In the amorphous semiconductor (or random network for a covalent-bond system), the structural defect is defined as an anomaly in the covalent bond with relation to the short-range order. Generally speaking, an atom of an element in the Nth group of the periodic table has N valence electrons, and when it is incorporated into a network, the network is energetically most stable when the coordination number is equal to N, if $N < 4$, or $(8 - N)$, if $N \geqslant 4$, according to the $(8 - N)$ rule of Mott. The bonding in such a network is called normal structural bonding (NSB) [42]. Any bonding which is different from NSB is defined as a defect, and such a configuration is known as a deviant electronic configuration (DEC) [42].

The five types of structural defect (or deviant electronic configurations) in a-Si:H that have been proposed up to now are illustrated in Fig. 3.16 in the form of two-dimensional models. Using Si with a coordination number of 4 as a standard (normal structural bonding (NSB)), the following bonds have been defined: dangling bonds for the undercoordinated Si atom of coordination number 3 [42], floating bonds for the overcoordinated Si atom of coordination 5 [43], and a three-centered bond for the case where a hydrogen atom is bound to two Si atoms [44]. If the definition of structural defects is extended, weak Si–Si bonds which have bond lengths greater than the normal Si–Si bond may also be regarded as another type of defect. In addition, undercoordinated Si atoms of coordination 2, charged defects (dangling bonds devoid of electrons, or containing two electrons), complex defects consisting of these defects coupled with nitrogen (N),

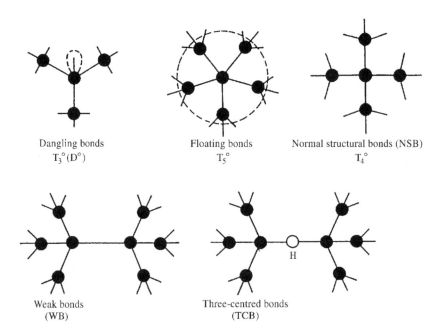

Fig. 3.16 Two-dimentional structural models illustrating the various types of defects (deviant electronic configurations, DECs) in a-Si:H; broken lines represent unpaired electrons. (see also Table 3.2)

carbon (C), or oxygen (O) included in a-Si:H, and impurity atoms such as donor atoms (P) or acceptor atoms (B), have been proposed [42, 45, 46]. For comparison purposes, various features of these structural models are summarized in Table 3.2. It should be noted that all of the proposed structural defects have not necessarily been demonstrated experimentally. Most of them, with the exception of T_3^0 and T_5^0, have been proposed for the purpose of understanding the characteristics of the localized levels within the band gap. This requires particular attention,

Table 3.2 Defect models proposed for amorphous silicon

Defect type (deviant electronic configuration)	Notation[a,b]	Remarks
Threefold coordinated Si (neutral)	T_3^0 (D^0)	Dangling bond (DB)
		Unpaired electron localized on Si s-component (closer to p^3) less than sp^3 Possible negative effective correlation energy? [46]
Threefold coordinated Si (positively charged)	T_3^+ (D^+)	No unpaired electrons [46]
		Close to sp^2?
Threefold coordinated Si (negatively charged)	T_3^- (D^-)	Non-bonding electron pairs
		Close to p^3?
Fivefold coordinated Si	T_5^0	Floating bond (FB) [43] Unpaired electron distributed over six Si atoms
Twofold coordinated Si	T_2^0	T_2^+, T_2^- in bandtail states? [46]; (positive effective correlation energy?)
Weak Si−Si bond	(WB)	Bandtail state? Antibonding states occur at localized tail states below the conduction band
Three-centered bond (complex defect with H)	(TCB)	Si−H−Si bonding; three electrons maintain two bonds Possible negative effective correlation energy? [44]
Complex defect with impurities	*D^- and others	$P_4^+ - T_3^-$ (*D^-), etc. [45] Complex defects with N, O, C, or B [42, 46] Negative effective correlation energy through lattice relaxation?
Normal structural bonding	T_4^0 (NSB)	Fourfold coordination in sp^3 hybrid orbital

[a]T = tetrahedrally bonded Si atom, where superscript represents the charged state and subscript the coordination number
[b]D = dangling bond

and will be discussed in detail below (see Chapter 4). It is very difficult to investigate directly the structure of the various defects by experimental methods, not only in the amorphous state but also in crystalline solids. Recent arguments concerning a form of defect in GaAs, known as 'EL2', nicely illustrate this situation [47].

3.5.2 ELECTRON SPIN RESONANCE SPECTROSCOPY STUDIES: DANGLING BONDS VERSUS FLOATING BONDS

The paramagnetic center detected through the use of electron spin resonance (ESR) spectroscopy provides, experimentally, the most detailed microscopic information concerning the structural defects in a-Si:H which involve unpaired electrons.

As with NMR spectroscopy, ESR spectroscopy is a magnetic resonance technique. The method does not detect covalent bonds based on pairs of electrons having upward and downward spins, and those defects which include similar two-electron states, but will detect, with great sensitivity, defects resulting from unpaired electrons which exhibit marked spin effects. When a magnetic field H is applied, the degenerated defect levels are split into two different energy states, i.e. upward-spin and downward-spin states. With a single electron and a vacant site, the value of the energy split is detected in the form of microwave resonance absorption. The Hamiltonian of the electron-spin system, H_{spin}, in a-Si:H is represented, using various major terms, by the following expression [48]:

$$H_{spin} = H_{ez} + H_{hfs} + H_{fs} \qquad (3.16)$$

The first term, H_{ez}, represents the electronic Zeeman interaction between the unpaired electron spin (of the spin Hamiltonian S) and the magnetic field H:

$$H_{ez} = \beta S \mathbf{g} H \qquad (3.17)$$

where β is the Bohr magneton ($\beta = e\hbar/2m_e C$), and $\mathbf{g}$ is known as the g tensor, which includes information related to the wave function and energy value of the unpaired electron. The second term in equation (3.16), H_{hfs}, represents the hyperfine interactions of the electron spin with spatially adjacent nuclear spins, while the third term, H_{fs}, represents the fine structure from the interaction between the symmetry of the electron-spin wave function and the surrounding crystal field. This last term comes into effect when electron spins are situated closely together.

In ESR spectroscopy, a sample is irradiated with microwaves from the X-band (around 9.2 GHz) or Q-band (around 35 GHz), and the magnetic field is scanned around the spin resonance energy to obtain absorption spectra with differentiated waveforms. The ESR spectrum obtained in this way includes the total contributions of all of the terms continued in equation (3.16). Fig. 3.17 shows a typical ESR spectrum of a-Si:H. The spectral waveform allows the determination of the resonance-center magnetic field (H_0), the spectral linewidth (ΔH_{pp}), and the

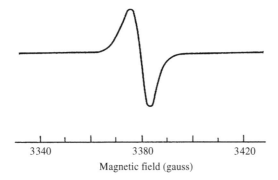

Fig. 3.17 A typical ESR spectrum of a-Si:H; measurements were made at room temperature, using the X-band (*ca.* 9.2 GHz)

spectral integral, which include information on the *g*-value (splitting constant), fluctuations in this parameter and in the local electric field, and the volume density of the unpaired electron spin N_S, respectively.

The resonance-center magnetic field H_0 corresponds to the transition energy between the energy levels $g\beta H_0/2$ and $-g\beta H_0/2$, which corresponds to Zeeman-split spins with magnetic moments of: $S = 1/2$ and $-1/2$, satisfying the following relationship:

$$h\nu = g\beta H_0, \tag{3.18}$$

where $\nu = 9.2$ GHz (X-band). From equation (3.18) it is possible to determine the *g*-value ($= h\nu/\beta H_0$) experimentally. For a-Si:H, we obtain $g = 2.0055$, which is larger than the *g*-value of the free electron ($= 2.0023$) [48, 49].

On the other hand, the values of ΔH_{pp} and N_S for a-Si are different from those of a-Si:H, as shown in Table 3.3. The most significant difference occurs in the spin density N_S, which ranges from 10^{18} cm^{-3} to 10^{20} cm^{-3} for a-Si and from 10^{15} cm^{-3} to 10^{17} cm^{-3} for a-Si:H, depending upon the method of preparation (see Table 3.3). The introduction of hydrogen markedly decreases the spin density. The FWHM also differs between a-Si and a-Si:H, with $\Delta H_{pp} \sim$ 7.5 G for a-Si:H, which is slightly larger than that found for a-Si, i.e. 5 G. The spectral linewidth (ΔH_{pp}) expands as the anisotropy of the *g*-value and the fluctuation in the local electric field grow, and the dipole interaction between the spins increases. If spin centers are brought so close together that the two wave functions overlap, then ΔH_{pp} is reduced as a result of exchange interaction, which is attributed to the contribution of the third term in equation (3.12), as observed in a-Si samples with $N_S > 10^{20}$ cm^{-3}. Detailed experimental studies have revealed the following: (1) ΔH_{pp} has little dependence on N_S in a-Si:H, (2) ΔH_{pp} of a-Si:H is nearly equal to that of a-Si, despite the presence of hydrogen at levels of tens of atom%, and (3) ΔH_{pp} of both a-Si and a-Si:H varies very little with changes in temperature. On the basis of these facts, the following conclusions may be drawn: (1) the spin centers in a-Si:H have their spin directions distributed

Table 3.3 Parameters characterizing the ESR spectrum of the paramagnetic center in a-Si:H and related materials

Spin center	g-Value	FWHM, ΔH_{pp} (G)	Spin density, N_S (cm^{-3})	Hyperfine splitting, ΔH_{hfs} (G)
a-Si:H	2.0055	~7.5	$10^{15}-10^{17}$	70 ± 10
P-doped	2.0043	–	–	–
B-doped	2.011	–	–	–
a-Si	2.0055	~4.5–5.0	$10^{18}-10^{20}$	130 ± 40
Si–SiO$_2$	2.0081	–	–	–
(P$_b$ center)	2.0012	–	–	–
c-Si				
(V$^+$)	2.0087	–	–	~40
	1.9989	–	–	–
(V$^-$)	2.0151	–	–	~125
	2.0028	–	–	–
	2.0038	–	–	–

randomly (powder pattern), and ΔH_{pp} is determined mainly by the anisotropy and fluctuation in the g-value, (2) the spin centers are isolated from one another, and the wave functions are strongly localized at the Si atoms, and (3) the spin centers are not spatially adjacent to the hydrogen [48, 50].

The above conclusions indicate that the dangling bond (DB) concept is a more appropriate representation of the defect model for the spin canters in a-Si:H than the floating bond (FB) concept. It is known that the P$_b$ centers at the Si/SiO$_2$ interface, which are considered to be composed of dangling bonds, have a mean g-value which is close to that of a-Si:H, and that ΔH_{pp}, calculated on the basis of a simulation of the powder pattern, is nearly equal to that of a-Si:H [48, 51].

Information concerning the wave function of the spin center is obtained from measurement of the ESR hyperfine structure which results from the ^{29}Si nuclear spin [52, 53]. This is attributed to the effect of the hyperfine interaction represented by the second term on the right-handside of equation (3.16). Since the natural abundance of ^{29}Si is only 4.7%, it is difficult to determine the hyperfine structure from ordinary ESR measurements because of the poor signal-to-noise (S/N) ratio. Some authors have reported hyperfine-structure split values of $\Delta H_{hfs} = 60$–70 G by special techniques, e.g. the use of using ^{29}Si-enriched samples by Biegelsen and Stutzmann [52], and the electron-nuclear double resonance (ENDOR) and pulsed ESR experiments (see Fig. 3.18) carried out by Yamasaki et al. [53, 54]. On the basis of these results, it may be claimed that while the charge density of the unpaired electron is localized at the Si atom, its value is as small as 0.05 at the center nucleus, and the 3s-orbital component of the wave function is 10%, which represents a coordination closer to p^3 rather than sp^3. Pantelides compared g-values and ΔH_{hfs} data with that obtained for the P$_b$ center and charged vacancies (V$^+$, V$^-$ in Table 3.3) and claimed that the spin center of a-Si:H was closer to V$^+$ with a broadened wave function (c-Si), thus

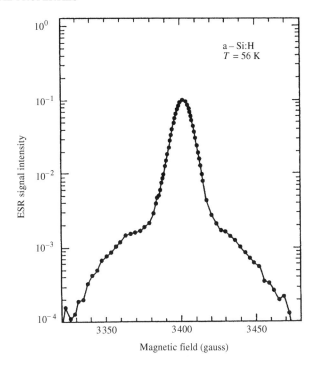

Fig. 3.18   ESR spectrum of a-Si:H obtained by using the pulsed ESR technique [53, 54, reproduced with permission. Copyright 1986, 1987 by the American Physical Society]

advocating the FB theory [55]. It should be pointed out that Pantelides' discussion is based on the idea that Floating bonds have reduced 3s-orbital components, in contradiction to the theoretical calculations carried out by Ishii and Shimizu [56].

The following discussion is based on the concept which attributes the electron-spin center of $g = 2.0055$ to the dangling bond (DB). N_S is the most important physical parameter for assessing the photoelectric features of a-Si-H, such as photoconductivity and photoluminescence, as will be discussed further in Chapter 4.

3.5.3 DEFECT DENSITY AND HYDROGEN IN FILMS

The conclusions reached in the preceding section are particularly important. While the ESR spin density N_S in a-Si, i.e. the concentration of dangling bonds, is 10^{20} cm^{-3} or so, it is reduced to *ca.* 10^{15} cm^{-3} in a-Si:H, owing to the inclusion of hydrogen in the network. This is the very reason why doping with impurities in a-Si:H is possible. However, the relationship of the dangling-bond concentration and hydrogen is not so straightforward.

Figure 3.19 shows plots the ESR spin density N_S, the total hydrogen content C_H (as determined by NMR spectroscopy) the broad-linewidth component of the

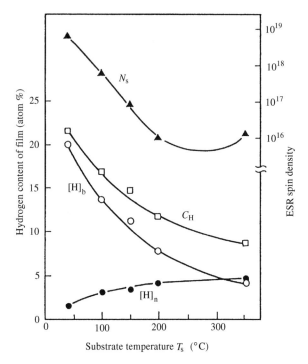

Fig. 3.19 The dependence of defect density (N_S), total hydrogen content C_H, concentrated hydrogen component $[H]_b$, and distributed hydrogen component $[H]_n$ on substrate temperature T_S [57, reproduced with permission]

hydrogen $[H]_b$ (concentrated hydrogen), and the narrow-linewidth component of the hydrogen $[H]_n$ (distributed hydrogen), against the substrate temperature (T_S) at the time of the a-Si:H preparation [57]. As is evident from this figure, the rule, "the higher the C_H, then the lower the N_S", does not always hold. It might be conceived that even if a large amount of hydrogen is contained, then N_S may remain at a high density level owing to insufficient structural relaxation at the surface at lower T_S values. In other words, in experiments, carried out with temperature as the parameter, the reduction in defects (N_S) due to thermal structural relaxation competes with the increase in defects caused by desorption of hydrogen. It does seem, however, that the correlation between the increase in the narrow-linewidth component $[H]_n$ and the decrease in N_S is fairly well established. Shimizu *et al.* studied the correlation between N_S and $[H]_n$ in a large number of samples of a-Si:H prepared both by the glow discharge and the reactive sputtering methods, and obtained the results shown in Fig. 3.20 [57]. While some of these samples did contain fluorine, it was demonstrated that the greater the $[H]_n$, then the lower N_S became in many of the samples. It seems that distributed hydrogen $[H]_n$, dispersed randomly within the films, effectively eliminates defects, but concentrated hydrogen, $[H]_b$, is not so effective.

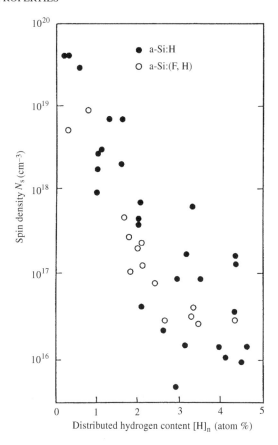

Fig. 3.20 ESR spin density versus distributed hydrogen content $[H]_n$ in various a-Si:H samples prepared by the glow discharge or reactive sputtering methods, including data obtained for a-Si:H:F ($\circ$) [57, reproduced with permission]

Figure 3.21 shows data obtained from both ESR and NMR spectroscopic measurements on a-Si:H samples, containing high concentrations of defects, which had been prepared with T_S held at room temperature, annealed at a fixed temperature for a certain period of time and the temperature then returned to ambient where the measurements were made [58]. The dependence of N_S, C_H, $[H]_b$, and $[H]_n$ on the annealing temperature T_a qualitatively resembles that observed for T_S shown in Fig. 3.19. As T_a rises, the total hydrogen content decreases, but the electron spin density also decreases as a result of thermal structural relaxation. However, $[H]_n$ only increases slightly within the correlation range shown in Fig. 3.20. When T_a is between 300 and 500 °C, the sample behaves as if $[H]_b$ has been partly converted to $[H]_n$ [58].

The above results suggest, in view of qualitative phenomenology, that the defect concentration is determined by (1) thermal effects and (2) the presence of

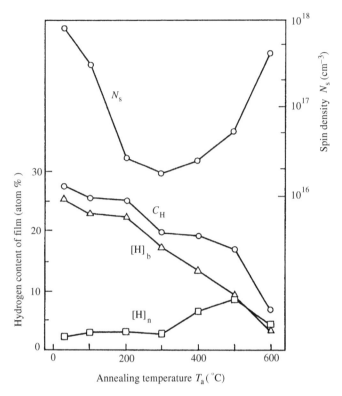

Fig. 3.21 Spin density (N_S), and hydrogen contents (C_H, $[H]_n$ and $[H]_b$) as functions of the annealing temperature of a-Si:H film samples prepared at room temperature [58, reproduced with permission]

hydrogen. These are the same conditions as those described in Section 2.3.4 for promoting structural relaxation on the surface of the material. The effect of the temperature T in equation (2.23) and the effect of E_S, reduced by the increase in hydrogen coverage of the surface, constitute the essential factors in the surface process model in which these effects enable the arriving chemical species to diffuse on the surface and thus ensure adequate structural relaxation. The data presented in Figs. 3.19–3.21 represent the structural properties of a-Si:H films formed through the surface process, and may be indirectly regarded as supporting the surface process model described in Chapter 2.

It should be noted, however, that while the defect concentration can be correlated with the distributed-hydrogen content, the one-to-one correspondence between the two fails in quantitative terms. The dangling bond (DB) concentration varies by 10^{16} cm^{-3}–10^{17} cm^{-3}, but the variation in hydrogen content is as small as a few atom%, i.e. in the order of 10^{20} cm^{-3}–10^{21} cm^{-3}. The situation is not so simple as to make a hydrogen binding to or being released from Si respectively correspond to decreasing or increasing the dangling bond concentrations. If the regular lattice network is fixed uniquely, as in case of a crystal, the one-to-one correspondence may be returned. For example, if it were possible to add hydrogen atoms one by one to each of the dangling bonds in a fixed network of a-Si containing 10^{20} dangling bonds per cm^3, then the defect density would in principle be reduced almost to zero with hydrogen levels of as little as 1 atom% or less. However, this is still a matter of discussion, and would be very difficult to implement it practically.

It is more important that seemingly a large quantity of hydrogen contained in actual a-Si:H is playing the roles of not only directly eliminating the dangling bonds, but also of modifying the properties of the network itself. According to Phillips' discussion of the structural stability described in Section 2.1.1, a-Si with a mean coordination number M of 4 has the form of an amorphous state based on covalent bonds with maximum strain energy. It is very difficult, therefore, to form an ideal amorphous structure without including defects, and the minimum structural relaxation is achieved by creating levels DB as high as 10^{20} cm^{-3}, i.e. by building atoms of coordination number 3. To extend this discussion, the realization of an amorphous structure in a-Si:H with DB levels as low as 10^{15} cm^{-3} may be attributed to the inclusion of hydrogen in amounts of several tens of atom % (atoms of single coordination), which reduces the mean coordination number M to 3.5–3.6, thus achieving adequate structural relaxation. In this way, it may be claimed that hydrogen serves as an agent for structural relaxation in the amorphous network [59].

To summarize, hydrogen plays the roles of (1) directly eliminating the number of dangling bonds, (2) controlling the surface processes in the formation of a-Si:H films, and (3) relaxing the network structures. These functions contribute indirectly in reducing the amount of dangling bonds, together with the thermal effects.

3.5.4 THERMAL EQUILIBRIUM PROCESS IN DEFECT FORMATION

The concentration of defects (i.e. dangling bonds) in a-Si:H is reduced by introducing hydrogen at a level of 10^{15} cm^{-3}, which corresponds to a concentration of 0.1 ppm, as the concentration of Si atoms is 5×10^{22} cm^{-3} in a-Si:H. This means that there is a dangling bond for every 10^6 Si atoms. Up until now, a large number of a-Si:H samples have been prepared by various methods and under several diverse conditions, but the lower limit of defect concentration has always been shown experimentally to be 10^{15} cm^{-3} for every type of sample. In principle, a-Si:H takes the form of highly diversified amorphous structures, and never has two identical structures. Nevertheless, the defect concentration at the lower limit can be accurately controlled to be as low as 0.1 ppm, irrespective of the preparative process. This is a surprising finding in view of the fact that thin films are prepared through non-equilibrium processes. We may also ask why is it impossible to reduce the defect concentration to below 10^{15} cm^{-3}?

As one attempt to try and answer these questions, the mechanism which decides the defect concentration has been examined through the concept of the thermal equilibrium process. This is similar to the method used in evaluating the defect concentration in crystalline materials (e.g. vacancies, interstitial atoms, Schottky defects and Frenkel defects) as a function of temperature under conditions where the free energy is minimized.

Let us examine two representative models, which still form the subject of active debate among specialists, and consequently have not yet been formulated in any definitive forms. It cannot be denied, however, that some of the non-equilibrium processes defining the defect concentration are of a nearly equilibrium nature. Street postulated a quasi-thermal equilibrium state on the surface of growing a-Si:H for the purpose of explaining the doping mechanism with phosphorus (P) (see Section 3.7). This was the first case of introducing the concept of equilibrium to a-Si:H, and opened the way to a number of subsequent arguments [60]. It should be noted, however, that the concentration of valence alternation pairs (VAPs) in chalcogenide glasses is evaluated by assuming that the equilibrium concentration at the glass transition temperature (T_g) is fixed [42]. This discussion is justified since a chalcogenide glass is frozen by passing through an equilibrium state considering of a supercooled liquid phase.

(a) Smith–Wagner model of equilibrium defect concentration

Smith and Wagner discussed quantitatively the defect (i.e. dangling bond) concentration of intrinsic (or undoped) a-Si:H on the basis of the conditions required for minimizing the free energy [61]. Let us assume that in a system containing N covalent bonds, M out of these N bonds are broken to generate $2M$ dangling bonds. If the ratio of the broken bonds to the total bonds is denoted by $\chi \equiv M/N(\chi \ll 1)$, and the entropy of mixing of the system by S, then the following expression is obtained by computing the number of possible arrangements of the randomly distributed dangling bonds (DBs):

$$S \cong -kN\chi \ln \chi \tag{3.19}$$

Therefore, the free energy $F(= H - TS)$ of the system can be written as follows:

$$F = \chi N U_B + kTN\chi \ln \chi \tag{3.20}$$

F is at a minimum when the following applies:

$$\chi^*(T) \cong \frac{1}{2N_0} \left[\frac{1}{2} N_d^*(T) \right] = e^{-1} \exp(-U_B/kT), \tag{3.21}$$

where k is the Boltzmann constant, U_B the energy added to the system by breaking a bond to create two dangling bonds, N_0 the concentration of Si atoms, and N_d the DB concentration produced by breaking the bonds. In other words, the DB concentration is defined so as to minimize the free energy of the system, with this concentration being known as the thermal-equilibrium defect concentration,

N_d^*. It should be noted, however, that as the amorphous network itself is assumed to be fixed, then the entire system is not in a thermodynamically steady state in the strictest sense. The problem, therefore, is how to calculate U_B.

In order to achieve this, the following assumptions are made: (1) U_B is defined as the difference between the electron energies of the two valence electrons constituting a covalent bond and that of the two DBs created by breaking this bond, and (2) if the energy axis is set within the energy gap (tail state of the valence band) with the origin placed at the mobility edge E_V of the valence band, then the density of tail states $N(E)$ is given as an exponential function of E, as follows:

$$N(E) = N_V \exp(-E/E_u) \tag{3.22}$$

where N_V is the density of states at $E_V = 0$, and is equal to 4×10^{21} cm^{-3} eV^{-1}, and E_u corresponds to the Urbach energy at the optical absorption edge, and is equal to ca. 50 meV. On the basis of these assumptions, covalent bonds are cut sequentially from those of higher energies in the valence-band tail state. This may be regarded as a process of forming DBs by cutting weak Si–Si bonds. The DB concentration reaches the thermal equilibrium concentration N_d^* when E is equal to E^* in the following expressions:

$$N_d^* = \int_{E*}^{\infty} N_V \exp(-E/E_u)dE = N_V E_u \exp(-E^*/E_u) \tag{3.23}$$

and $E^* = E_u \ln(N_V E_u / N_d^*)$ $\tag{3.24}$

In consideration of the fact that the energy position of the dangling bond is $E_d (E^* < E_d)$ and that the density of the tail states $N(E)$ varies exponentially, N_d^* can be represented by a value within a range $E^* \pm E_u$ (see Fig. 3.22). We finally arive at the following expressions for U_B and N_d^* [61]:

$$U_B \cong 2E_d - 2E^* \tag{3.25}$$

and

$$N_d^* = [4N_0 e^{-1} \exp(-2E_d/kT)]^{kT/(2E_u+kT)} \times [N_V E_u]^{2E_u/(kT+2E_u)} \tag{3.26}$$

While equation (3.26) is somewhat complicated in its present form, the important parameters are the energy required to form the dangling bonds, E_d, and the slope of the exponential tail for the valence band, E_u. The expected temperature dependence of N_d^* is shown in Fig. 3.23. When $E_d = 0.85$ eV and $E_u = 50$ meV, the DB concentration of a-Si:H film grown at a substrate temperature T_S of 520 K becomes $N_d^* \cong 10^{16}$ cm^{-3}, and if this concentration is frozen in, then the actual value would be more closely approached.

The slower the quenching rate, and the faster the rate that the network reaches thermal equilibriums, then the lower the freeze-in temperature will become (see

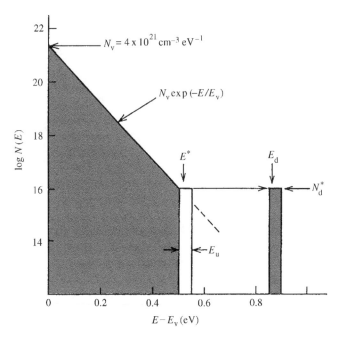

Fig. 3.22 Illustration of the use of the Smith–Wagner model to determine the thermal equilibrium defect concentration N_d^* [61, reproduced with permission. Copyright 1987 by the American Physical Society]

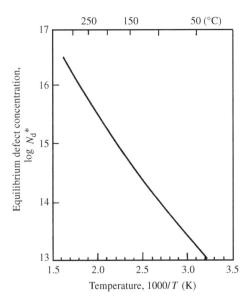

Fig. 3.23 Temperature dependence of the thermal equilibrium defect concentration N_d^* [61, reproduced with permission. Copyright 1987 by the American Physical Society]

Chapter 1). Smith and Wagner claimed that the equilibrating rate depends upon the diffusion rate of hydrogen within the network, which is close to the 'hydrogen glass' model of Street *et al.* [62] (see Section 5.4 below). The temperature at which thermally moving hydrogen atoms occur, which are responsible for the defect concentration equilibrium, is defined as the glass transition temperature. If no hydrogen exists, then equilibrium is not reached and the defect concentration is frozen in at a higher level.

However, the hypothesis of Smith and Wagner has not yet been fully verified experimentally, although this model has successfully explained for the first time why it is difficult to reduce N_d to below 10^{15} cm^{-3}. When this model can be applied, it is essential to reduce N_d to create an a-Si:H network of smaller E_u (i.e. with the exponential tail of the valence band sloping steeply down) and to accelerate the rate for attaining equilibrium.

(b) Winer's chemical equilibrium model

Winer reported a model which considers hydrogen in greater detail [63]. While the Smith–Wagner model is based on the concept of the generation of Schottky defects in the crystal, the Winer model is closer to the concept of Frenkel defects in which both vacancies and interstitial atoms are created simultaneously (see Fig. 3.24). In this model, the process of defect generation is regarded to be as shown in Fig. 3.24(b) in analogy to that which takes place in the crystal (see Fig. 3.24(a), namely:

$$\text{SiH} + \text{Si–Si} \rightleftharpoons \text{D}° + \text{SiHSi} \tag{3.27}$$

where SiH is assumed to exist at a concentration close to that of hydrogen, Si–Si is a weakbond forming the tail state of the valence band, and SiHSi represents H captured at the weak-bond site, constituting an ESR center which is not distinguishable from an isolated dangling bond (D°). This should be regarded as being quite different from the three-centered bond discussed earlier (see Fig. 3.16 and Table 3.2). Although the calculation is not detailed here, equation (3.27) can be regarded as representing a chemical reaction, and the defect concentration is determined on the basis of the conditions for chemical equilibrium. In this case, it is assumed that (1) the defect states (D°, D$^+$ and D$^-$) are distributed with respect to energy as a Gaussian-type function and the correlation energy is positive, and (2) the charged state of the defects depends upon the Fermi energy E_F, and the energy position of the defect is defined on the basis of an equilibrium state (with respect to the chemical potential). The defect concentration, determined as a function of E_F, is 10^{15} cm^{-3} for undoped a-Si:H, which will be further increased if doped to give p- or n-type materials (Fig. 3.25).

Both the Smith–Wagner and Winer representations are to be regarded simply as models. For example, the close association of the DB with H, as shown in Fig. 3.24, has not yet been verified experimentally, and there are even data which denies such a correspondence. The two models are only meaningful from

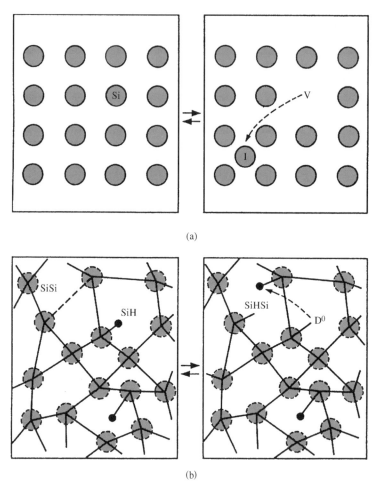

Fig. 3.24 (a) Generation of Frenkel defect in c-Si. (b) Generation of defects in a-Si:H based on the reactionship represented by equation (3.27) [63, reproduced with permission. Copyright 1990 by the American Physical Society]

the viewpoint of attempting to explain how the defect concentration in a-Si:H is determined through some type of equilibrium process. If a new non-equilibrium process were found to which these models could not be applied, it might, in principle, be possible to break through the barrier of $N_d = 10^{15}$ cm^{-3}. It should be noted that these two models do not preclude this possibility.

In fact, Ganguly and Matsuda in recent experiments, have achieved defect concentrations of the order of 10^{14} cm^{-3} on the basis of a surface diffusion model, which is quite different from the equilibrium model [64]. This model is an extension of the surface process studies described in Chapter 2, and is expected to play a central role in future dicussions.

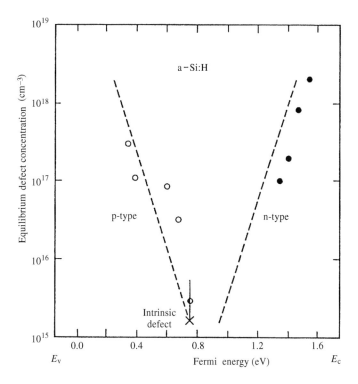

Fig. 3.25 Dependence of equilibrium defect concentration on Fermi energy, showing calculated data (broken lines and ×), together with experimental data based on activation energies for the conductivity and photothermal deflection spectroscopy (PDS) using B-doped (○) and P-doped (●) samples [63, reproduced with permission. Copyright 1990 by the American Physical Society]

3.6 Impurities

Impurities in this part of the discussion do not refer to so-called dopant impurities, i.e. donors and acceptors, which are intentionally added for controlling the valence electrons, but to those impurity atoms which are taken into a-Si:H as uninvited guests during the process of preparation.

As stated previously, the concentration of defects (i.e. dangling bonds), which can be measured as the spin density in ESR spectroscopy, is as low as 10^{15} cm^{-3} in a-Si:H. Nevertheless, there are impurity atoms of some species which are contained in a-Si:H at a level which is higher than this. Representative impurity species include oxygen, nitrogen and carbon, which may attain to concentrations as high as 10^{18} cm^{-3} – 10^{20} cm^{-3} providing that the samples are prepared by the ordinary glow discharge method [4].

According to Carlson [65], oxygen may be admitted either from the walls of the vacuum chamber in the form of H_2O, CO or CO_2, or as a contaminant of the SiH_2 source gas in the form of $(SiH_3)_2O$, nitrogen may come via leakage of air into the vacuum chamber, release from the chamber walls or contaminant N in the SiH_4 cylinder gas, and carbon may be introduced as CO or CO_2 in the exhaust gas or as back-diffused hydrocarbons from the vacuum pump.

In order to reduce these impurities, Tsai et al. [66] and Nakano et al. [67] have attempted plasma CVD experiments based on a ultrahigh vacuum (UHV) system. Tsai and coworkers succeeded in reducing the impurity levels down to [O] = 3×10^{17} cm^{-3}–2×10^{18} cm^{-3}, [C] = 2×10^{17} cm^{-3}, [N] = 6×10^{16}–9×10^{17} cm^{-3} and [Cl] = 10^{17} cm^{-3}, which are still higher than the dangling bond concentration. Later, the Mitsui–Toatsu group [68] successfully lowered the nitrogen level to 10^{15} cm^{-3}.

No systematic study has yet been made on how these highlevel of impurities are incorporated into the network, and what effects they exert on the defect generation and transport phenomenon.

3.7 Doping

The preceding sections have concerned the structural properties of undoped a-Si:H. Let us now examine the structure of a-Si:H with the conductivity controlled to be either n-type or p-type by doping with phosphorus (P) or boron(B), respectively. As described in Chapter 2, n-type a-Si:H is prepared by glow discharge of a PH_3/SiH_4 mixed gas (often diluted with Ar or H_2), and p-type a-Si:H from B_2H_6/SiH_4 mixtures. The doping characteristics of phosphorous and boron, the hydrogen contents of the films, and results obtained from ESR spectroscopy, will be outlined below.

3.7.1 DOPING CHARACTERISTICS AND EFFICIENCIES: PHOSPHORUS AND BORON

In preparations based on the glow discharge technique, the atomic ratios in the solid phase of a-Si:H, i.e. [P]/[Si] and [B]/[Si], are essentially proportional to the mixed gas ratios, $[PH_3]/[SiH_4]$ and $[B_2H_5]/[SiH_4]$, respectively. In quantitative comparisons [69], the atomic ratio in the gas phase is equal to that in the solid phase within one order of magnitude over a range of atomic ratios from 10^{-5} to 10^{-2}.

It is questionable, however, whether or not P or B are incorporated into a-Si:H as a fourfold coordinated atoms and operate as donors or acceptors, respectively. As mentioned in Chapter 1, Mott has predicted that atoms of any species will be incorporated into an amorphous network so as to meet the valence electron requirements. Therefore, according to the $(8 - N)$ rule of Mott, both phosphorus (group V in the periodic table) and boron (group III) are energetically stable if they are threefold coordinated, in which case both of these atoms can be neither donor nor acceptor. The possibility of impurity control (pn control) depends upon whether a dopant atom becomes three- or four-fold coordinated. Earlier, Knights et al. added arsenic (As, group V), in place of phosphorus, to a-Si:H through the glow discharge of SiH_4/AsH_3, and determined the extended X-ray absorption fine structure (EXAFS) of As [70]. According to their data, it may be concluded that *ca.* 20% of As in the solid phase, of concentration 4×10^{20} cm^{-3}, is fourfold coordinated [70]. However, this value is one or more orders of magnitude higher than the doping efficiency obtained from subsequent experiments, and it is speculated that hydrogen in the network broadly affected the EXAFS analysis [71]. Hayashi *et al.* report, on the basis of NMR spectroscopy using ^{31}P, that most of the P atoms in the solid phase have three-fold coordination and do not form bonds with hydrogen [72].

As will be detailed in Chapter 4, it is known that the concentration of donor (fourfold coordinated P), N_D, is proportional to the square root of the total P concentration, N_{P0}, in the solid phase [73]. In order to explain this, Street introduced the concept of metastable thermal equilibrium at the growing surface, while applying the $(8 - N)$ rule of Mott [73]. As shown in Fig. 3.26, the added P atoms are allowed to exist as ionized donors (P_4^+), but not in the state filled with electrons (P_4^0). Since P_4^0 has five valence electrons in the outermost shell, its coordination number should be $8 - 5 = 3$, according to the $(8 - N)$ rule of Mott, and hence, P_3^0 (threefold coordination) is the most stable. In a similar way, the deep tail states of the conduction band below E_F are occupied by electrons. Therefore, the T_4 state has five valence electrons, and so, if the $(8 - N)$ rule is applied by assuming electron localization, the stable state is not T_4^- but T_3^- in threefold coordination, i.e. the dangling bond, D^-. For added P atoms, the following reaction occurs depending upon the position of the Fermi level E_F,

$$P_3^0 \rightleftharpoons P_4^+ + D^- \tag{3.28}$$

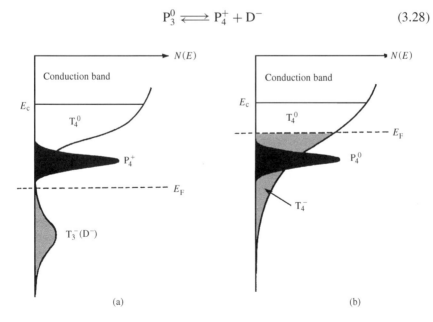

(a)   (b)

Fig. 3.26 Schematic diagrams showing the density of states in n-type doping of a-Si:H with phosphorus: (a) with the $(8 - N)$ rule applied (Street's model); (b) with the $(8 - N)$ rule not considered (if considered, P_4^0 and T_4^- become P_3^0 and T_3^-, respectively)

In this way, whenever a P_4^+ of fourfold coordination (donor) is formed, a DB is created as if to compensate it. For this reason, E_F cannot extend to the shallow tail states. The following expression is derived from the law of mass action:

$$(N_{P0} - N_D)/N_D^2 = \text{constant} \tag{3.29}$$

If the doping efficiency η in the solid phase is defined by equation (3.30), then the relationship (3.31) is obtained from equation (3.29) since $\eta \ll 1$ (or $N_D \ll N_{P0}$) in a-Si:H as stated above:

$$\eta = \frac{N_D}{N_{P0}} \tag{3.30}$$

$$\eta = \frac{N_D}{N_{P0}} \cong \text{constant}/N_{P0}^{1/2}, \text{ or } N_D \propto N_{P0}^{1/2} \tag{3.31}$$

which will be helpful for explaining the experimental data [73].

The model of Street described in the above is characterized by his original assumption that a quasi-thermal equilibrium process occurs at the surface when a-Si:H is formed from a $(SiH_4 + PH_3)$ plasma. More detailed investigations, which followed have confirmed that the dependency of the doping efficiency η on N_{P0} is applicable not only for P, but also for B, As and even doping in a-Ge:H, satisfying equation (3.31). (see later in Fig. 4.23, in Section 4.3.4) [71].

3.7.2 HYDROGEN IN DOPED a-Si:H

Data for the content and bonding configurations of hydrogen in a-Si:H thin films of both p-type (doped with B) and n-type (doped with P) are available as measured by the hydrogen evolution and infrared absorption methods described in Section 3.2 [69]. In comparison to CVD from a SiH_4 plasma, more complicated chemical and surface processes are involved in deposition when using $(SiH_4 + PH_3)$ or $(SiH_4 + B_2H_6)$ plasmas, thus causing quite different modes of hydrogen inclusion between P-doped and B-doped films.

The total content of hydrogen, C_{TH} determined from hydrogen thermal evolution curve $(H(T))$, where:

$$C_{TH} \propto \int H(T)dT$$

is nearly constant at $C_{TH} \cong 15$ atom%, as long as the mixed gas ratio $[PH_3]/[SiH_4]$, which is proportional to the doping level $[P]/[Si]$, ranges from 10^{-5} to 10^{-3}, in the case of P-doped samples [69]. This value agrees fairly well with the value of the total bonded-hydrogen content C_H, as determined by infrared absorption, and is similar to the case of intrinsic or undoped a-Si:H. However, the situation is quite different for B-doped samples. As the mixed gas ratio $[B_2H_6]/[SiH_4]$ rises, the hydrogen evolution C_{TH} increases from 15 to nearly 25 atom%, while the total bonded-hydrogen content C_H decreases from 15 to just a few atom%. It seems that the more the film is doped with B, then the less the hydrogen which is bonded with Si and the more that the molecular hydrogen (H_2) increases. This nicely corresponds to the marked deterioration in various macroscopic aspects of the films, as observed by using TEM, with increases in the B doping levels [23]. Hydrogen in doped a-Si:H diffuses more readily through the network (for the diffusion coefficient, refer to Fig. 5.13 in Chapter 5.4).

As the bonded-hydrogen content is reduced, the optical bandgap decreases. Studies on B-doped a-SiC:H are in progress, with carbon (C) being introduced as a means of preventing the bandgap from being reduced, while still retaining high film quality, as will be described later in Chapters 4 and 6.

3.7.3 ELECTRON SPIN RESONANCE STUDIES

The ESR signal at $g = 2.0055$ observed in undoped a-Si:H is identified as originating from the dangling bond. As the Fermi level moves to the band edge as a result of doping, different ESR signals will appear: a signal of $g = 2.0055$ is replaced by one with $g = 2.0044$ as P-doping is augmented, and by one with $g = 2.011$ as B-doping advances [49]. Some authors have claimed that the signals where g is 2.0044 and 2.011 originate from electrons trapped in the tail state of the conduction band, and from positive holes trapped in the tail state of the valence band, respectively [49, 71]. However, recent measurements obtained by using the latest pulse ESR technique suggest that they originate from localized centers close to dangling bonds. It is hoped that future work will decide which explanation is correct [54].

In addition to these observations, electron spins (9.14 GHz, $H_0 = 3.255$ G, $\Delta H_{hfs} = 240$–245 G; $g = 2.003$) which are strongly coupled with the nuclear spins of phosphorus ($^{31}P : I = 1/2$) have been reported [69, 71]. These signals are attributed to the hyperfine interaction represented by the second term on the right-handside of equation (3.16), and according to the analysis by Stutzmann $et\ al.$, this electron spin center is to be regarded as the donor center, and its size is much smaller (around 10 ± 1 Å), than the Bohr radius (16.7 Å) of the wave function for the donor (P atom) in crystalline silicon [71]. However, Yamasaki and Tanaka have reported contrary evidence to the concept of a donor center, with the origin of this signal having not yet been defined [69]. On the other hand, P-doped samples give only a signal of $g = 2.011$ ($\Delta H_{pp} = 24$ G), with the existence of electron spins indicating hyperfine interactions with B atoms (10B, 11B) not yet confirmed [69].

Signals with $g = 2.0044$ and 2.011 correspond to the light-induced ESR (LESR) signals observed in intrinsic (non-doped) a-Si:H at temperatures lower than 150 K under illumination [74]. Phenomenologically, this effect can be explained by the shift of the quasi-Fermi level to the neighbourhood of the band edges under illumination by light (see later in Section 5.2.2).

References

1. A. Bienenstock, *Amorphous and Liquid Semiconductors*, edited by J. Stuke and W. Brenig, Taylor & Francis, London (1973) 49.
2. J. C. Knights, *Jpn J. Appl. Phys.*, **18** (1979) Suppl. 1, 101.
3. A. J. Leadbetter, A. A. M. Rashid, R. M. Richardson, A. F. Wright and J. C. Knights, *Solid State Commun.*, **33** (1980) 973.
4. K. Tanaka and A. Matsuda, *Mater. Sci. Rep.*, **2** (1987) 139.
5. T. Ichimura, T. Ihara, T. Hama, M. Ohssawa, H. Sakai and Y. Uchida, *Jpn J. Appl. Phys.*, **25** (1986) L 276.
6. G. Lucovsky and W. B. Pollard, *The Physics of Hydrogenated Amorphous Silicon, Part II*, Topics in Applied Physics, **56**, Springer, Berlin (1984) 301.
7. C. W. Magee and D. E. Carlson, *Solar Cells*, **2** (1980) 365.
8. W. A. Lanford, H. P. Trantuetter, J. F. Ziegler and J. Keller, *Appl. Phys. Lett.*, **30** (1977) 566.
9. H. Fritzsche, M. Tanielan, C. C. Tsai and P. J. Gaczi, *J. Appl. Phys.*, **50** (1979) 3366.
10. S. Yamasaki, S. Kuroda, K. Tanaka and S. Hayashi, *Solid State Commun.*, **50** (1984) 9.
11. J. A. Reimer, *J. Phys. (Paris)*, **42** (1981) C 4-715.
12. W. Beyer, *Tetrahedrally Bonded Amorphous Semiconductors*, edited by D. Adler and H. Fritzsche, Plenum Press, New York, (1985) 129.
13. W. Paul and D. A. Anderson, *Solar Energy Mater.*, **3** (1981) 229.
14. K. Tanaka, *Applied Physics Handbook*, edited by the Japanese Society of Applied Physics, Maruzen, Tokyo (1990) 603 (in Japanese).
15. C. J. Fang, K. J. Gruntz, L. Ley and M. Cardona, *J. Non-Cryst. Solids*, **35/36** (1980) 255.
16. H. Shanks, C. J. Fang, L. Ley, M. Cardona, F. J. Demond and S. Kalbitzer, *Phys. Status Solids(b)*, **100** (1980) 43.
17. M. Cardona, *Phys. Status Solids(b)*, **118** (1983) 463.
18. W. Beyer and H. Wagner, *J. Non-Cryst. Solids*, **59/60** (1983) 161.
19. W. Beyer and H. Wagner, *J. Appl. Phys.*, **53** (1982) 8745.
20. D. E. Polk, *J. Non-Cryst. Solids*, **5** (1971) 365.
21. M. H. Brodsky, M. H. Kaplan and J. F. Ziegler, *Appl. Phys. Lett.*, **21** (1972) 305.
22. M. H. Brodsky, M. A. Frisch, J. F. Ziegler and W. A. Lanford, *Appl. Phys. Lett.*, **30** (1977) 561.
23. H. Fritzsche, *Solar Energy Mater.* **3** (1980) 447.
24. K. Tsuji and S. Minomura, *J. Phys. (Paris)*, **42** (1981) C4-233.
25. P. John, *J. Phys. C: Solid-State Phys.*, **14** (1981) 309.
26. D. Weaire, N. Higgins, P. Moore and I. Marshall, *Phil. Mag.*, **B 40** (1979) 243.
27. F. Yonezawa, *Fundamentals of Amorphous Semiconductors*, edited by M. Kikuchi and K. Tanaka, OHM Publishing Company, Tokyo (1982), p. 29 (in Japanese).
28. J. S. Lannin, *Phys. Today* (July 1988) 28.
29. S. C. Moss and J. F. Graczyk, *Proceedings of the 10th International Conference on the Physics of Semiconductors*, edited by S. P. Keller, United States Atomic Energy Commission, Oak Ridge (1970) 658.
30. F. Wooten, K. Winer and D. Weaire, *Phys. Rev. Lett.*, **54** (1985) 1392.
31. N. F. Mott and E. A. Davis, *Electronic Processes in Non-Crystalline Materials*, 2nd Edn, Clarendon Press, Oxford, (1979) 322.
32. G. A. N. Connell and R. J. Temkin, *Phys. Rev.*, **B 9** (1974) 5323.
33. D. E. Polk and D. S. Bondreaux, *Phys. Rev. Lett.*, **31** (1973) 92.
34. R. Mosseri, J. C. Mclaurent, C. Sella and J. Dixmier, *J. Non-Cryst. Solids*, **35/36** (1980) 507.

35. R. Bellissent, A. Chenevas-Paule, P. Cheiux and A. Menelle, *J. Non-Cryst. Solids*, **77/78** (1985) 213.
36. T. Ishidate, K. Inoue, K. Tsuji and S. Minomura, *Solid State Commun.*, **42** (1982) 197.
37. D. Beeman, R. Tsu and M. F. Thorpe, *Phys. Rev.*, **B 32** (1985) 874.
38. R. Shuker and R. W. Gammon, *Phys. Rev. Lett.*, **25** (1970) 222.
39. R. Bellissent, *Amorphous Silicon and Related Materials: Volume A*, edited by H. Fritzsche, World Scientific, (1989), p. 93.
40. S. Muramatsu, S. Matsubara, T. Watanabe, T. Shimada, T. Kamiyama, K. Suzuki and A. Matsuda, *Jpn J. Appl. Phys. Lett.*, **30** (1991) L 2006.
41. Z. E. Smith, *'Glow-Discharge Hydrogenated Amorphous Silicon'* edited by K. Tanaka, KTK/Kluwer, Tokyo and Boston, (1989), p. 101.
42. T. Shimizu, *Fundamentals of Amorphous Semiconductors*, edited by M. Kikuchi and K .Tanaka, OHM Publishing Company, Tokyo (1982) 59, (in Japanese).
43. S. T. Pantelides, *Phys. Rev. Lett.*, **57** (1986) 2979.
44. R. Fisch and D. C. Licciardello, *Phys. Rev. Lett.*, **41** (1978) 889.
45. K. Tanaka, H. Okushi and S. Yamasaki, *Tetrahedrally Bonded Amorphous Semiconductors*, edited by D. Adler and H. Fritzsche, Plenum, NewYork (1985) 239.
46. D. Adler, *Phys. Rev. Lett.*, **41** (1978) 1755; *J. Phys. (Paris)*, **42** (1981) C4-3.
47. A. Yawata, *Oyo-Buturi*, **57** (1988), 87 (in Japanese).
48. P. C. Taylor, *Hydrogenated Amorphous Silicon, Part C: (Semiconductors and Semimetals*, edited by J. I. Pankove, Academic Press, NY, **21** (1984) 99.
49. H. Dersch, J. Stuke and J. Beichler, *Phys. Status Solids(b)*, **105** (1981) 265.
50. P. A. Thomas and D. Kaplan, *Structure and Excitations of Amorphous Solids*, edited by G. Lucovsky and F. L. Galeener, American Institute of Physics, NY, (AIP Conference Proceedings No.31. (1976) 85.
51. K. L. Brower, *Appl. Phys. Lett.*, **43** (1983) 1111.
52. K. Biegelsen and M. Stutzmann : *Phys. Rev.*, **B 33** (1986) 3006.
53. S. Yamasaki, M. Kaneiwa, S. Kuroda, H. Okushi and K. Tanaka, *Phys. Rev.*, **B 35** (1987) 6471.
54. S. Yamasaki, *et al.*, *Phys. Rev. Lett.*, **65** (1990) 756.
55. S. T. Pantelides, *Phys. Rev. Lett.*, **57** (1986) 2979.
56. N. Ishii and T. Shimizu, *Jpn J. Appl. Phys.*, **27** (1988) L1800.
57. T. Shimizu, K. Nakazawa, M. Kumeda and S. Ueda, *Physica*, **117 B/118 B** (1983) 926.
58. S. Yamasaki, S. Kuroda, K. Tanaka and S. Hayashi, *Solid State Commun.*, **50** (1984) 9.
59. T. Shimizu: *Oyo-Buturi*, **56** (1987), 900 [in Japanese].
60. R. A. Street, *Phys. Rev. Lett.*, **49** (1982) 1187.
61. Z. E. Smith and S. Wagner, *Phys. Rev. Lett.*, **59** (1987) 688.
62. R. A. Street, J. Kakalios, C. C. Tsai and T. M. Hayes, *Phys. Rev.*, **B 35** (1987) 1316.
63. K. Winer, *Phys. Rev.*, **B 41** (1990) 12150.
64. G. Ganguly and A. Matsuda, *Jpn J. Appl. Phys. Lett.*, **31** (1992) L1269.
65. D. E. Carlson, *Hydrogenated Amorphous Silicon, Part D: Semiconductors and Semimetals*, edited by J. I. Pankove, Academic Press, NY, **21**, (1984) 1.
66. C. C. Tsai, J. C. Knights, R. A. Lujan, B. Wacker, B. L. Stafford and M. J. Thompson, *J. Non-Cryst. Solids*, **59/60** (1983) 731.
67. S. Nakano, S. Tsuda, H. Tarui, T. Tanakawa, H. Haku, K. Watanabe, M. Nishikuni, Y. Hishikawa and Y. Kuwano, *Matter. Res. Soc.*, **70** (1986) 511.
68. T. Kuriha, *Proceedings of the 50th Meeting of the Japanese society of Applied Physics*. No. 2 (1989) 709 (in Japanese).
69. S. Yamasaki and K. Tanaka, *Glow-Discharge Hydrogenated Amorphous Silicon*, edited by K. Tanaka. KTK/Kluwer, Tokyo and Boston (1989) 127.

70. J. C. Knights, T. M. Hayes and J. C. Mikkelsen, Jr., *Phys. Rev. Lett.*, **39** (1977) 712.
71. M. Stutzmann, D. K. Biegelsen and R. A. Street, *Phys. Rev.*, **B 35** (1987) 5666.
72. S. Hayashi, K. Hayamizu, S. Yamasaki, A. Matsuda and K. Tanaka, *Phys. Rev.*, **B 38** (1988) 31.
73. R. A. Street, *Phys. Rev. Lett.*, **49** (1982) 1187.
74. R. A. Street and D. K. Biegelsen, *J. Non-Cryst. Solids*, **35/36** (1980) 651.

4 OPTICAL
AND ELECTRICAL
PROPERTIES

This chapter will concern itself with detailed descriptions of the basic concepts of the optical and electrical properties of amorphous semiconductors with respect to the practical example of amorphous silicon. Using a simplified physical model, we will discuss how the electronic states in the vicinity of the band edge, and the optical and electric transport processes associated with these, are affected by the structural disorder. Attempts will be made to interpret and explain the various physical properties of amorphous silicon on both the basis of this discussion and the structural-defect model described in Chapter 3.

4.1 Disorder and Electronic Structure

4.1.1 ELECTRONIC STRUCTURE OF AMORPHOUS SEMICONDUCTORS

In a covalently bonded solid, such as amorphous silicon, the short-range order (SRO), equivalent to the corresponding crystal line material, is well preserved in its local chemical bond configuration or coordination number of constituent atoms, despite its designation 'amorphous'. Under such a situation, where SRO exists and all of the constituent atoms form a continuous network, it has been precisely demonstrated, by using a simple tight-binding model Hamiltonian, that the energy gap occurs just as in the crystal [1], as earlier described in Section 1.2.1.

On the basis of this conclusion, an intuitive argument might proceed in the following way. In a-Si:H, each of the Si atoms is bonded to the four other Si atoms surrounding it by four chemical bonds based on sp^3 hybrid orbitals, and locates itself at the center of gravity of a tetrahedron formed by the four neighbouring Si atoms. Here, eight hybrid orbitals are separated into four bonding and four antibonding orbitals. The bonding orbitals of lower energy are filled by four outer valence electrons supplied from each of the Si atoms, while the bonding orbitals remain empty. As a result of interaction between the adjacent orbitals, the bonding orbitals form a valence band filled with electrons, while the antibonding orbitals form an empty conduction band, as schematically illustrated in Fig. 4.1. A finite

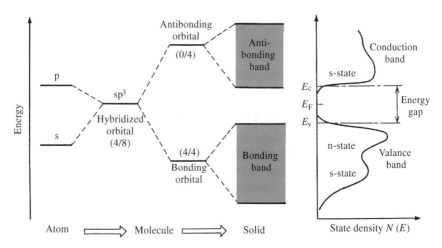

Fig. 4.1 Band structure of a tetrahedrally bonded semiconductor as portrayed by the use of chemical bond theory

energy gap exists between the two bands, and as the Fermi level is located within the gap, a-Si shows the properties of a semiconductor.

While this discussion applies to both the amorphous and crystalline states, there is a distinct difference between the electronic structures of the two states. Before continuing our discussion, the following point needs to be taken into consideration. In a crystalline semiconductor with long-range order, the wave vector k of the electron is a 'good' quantum number for describing the electronic state on the basis of the Bloch theorem, and a so-called dispersion relationship, $E = E(k)$, exists between the electron energy E and the wave vector k. The discussion on the electronic structure of the crystal semiconductor depends on the basis of this dispersion relationship. In the case of the amorphous semiconductor, which lacks long-range order, the wave vector is no longer a 'good' quantum number, and the dispersion relationship $E = E(k)$ cannot be defined. In its place, the energy density-of-states (DOS) distribution function, $N(E)$, becomes the major subject of the discussion of the electronic structure of amorphous semiconductors.

The DOS spectra of a crystal semiconductor has a specific structure (the van Hove singularity) reflecting the periodicity and symmetry of the atomic array, which naturally cannot exist in the amorphous semiconductor. Moreover, owing to short-range structural disorder, such as bond length and bond angle, as well as the fluctuation in the intermediate-range structure, such as dihedral angle and ring topology, the DOS of the amorphous semiconductor takes a blurred form of the DOS which corresponds to that of the crystal [2]. The effects of such a structural disorder are clearly recognized at the band edge, where the latter strongly affects the electronic processes in the semiconductor, i.e. the sharp band edge in the crystal is tailed into the energy gap, thus taking the form of a localized band tail.

If the structural disorder in a-Si is caused by relaxing the internal stress of the network made up from an overconstrained tetrahedral bonding structure in order to stabilize the overall structure, it is no wonder that a-Si includes various structural defects such as orbitals not involved in bonding (dangling bonds) for these same reasons. Most of the structural defects, caused by both stress relaxation and for thermodynamic reasons, are thought to form levels localized within the energy gap [3].

4.1.2 CONDUCTION BAND AND VALENCE BAND

Following the above general introductory, the band states, the band tail states and the structural defects in a-Si:H will be treated in sequence. Fig. 4.2 shows

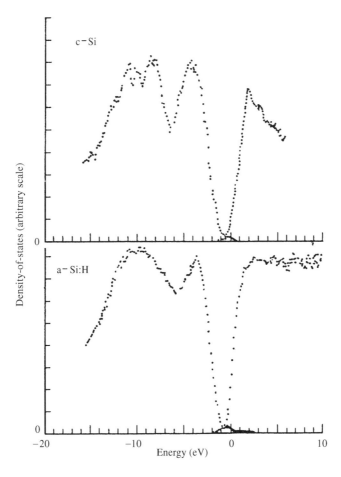

Fig. 4.2 The density-of-states spectra of c-Si and a-Si:H recorded by photoemission spectroscopy without any corrective measures [4, reproduced with permission. Copyright 1985 by the American Physical Society]

the DOS spectra for the valence and conduction bands in a-Si:H and c-Si, as recorded by X-ray photoemission spectroscopy (XPS) and inverse photoemission spectroscopy [4]. While the DOS spectrum for the amorphous state is blurred in comparison to that of the crystalline state, it is evident that the two closely resemble each other. This directly demonstrates that the short-range order, and the electronic structure based on it, in a-Si:H are almost identical to those in c-Si in a general sense.

The valence-band spectrum of c-Si has three peaks, which can be identified, on the basis of theoretical calculations for a cluster model [5], as being attributable to a p orbital, an sp hybrid orbital and an s orbital, from the top of the valence band [6]. In a-Si:H, two peaks on the lower-energy side are fused to form a broader structure. While this has been previously ascribed to the presence of five-membered ring structures in the Si network, it has also been shown that the same effects can be caused by quantitative disorder such as fluctuations in the dihedral angle [8, 9].

As a-Si:H contains 5–20 atom% of bonded hydrogen, its effect on the DOS spectrum is not to be ignored. In the XPS experiments mentioned above, the component associated with hydrogen, i.e. Si–H_x, is not explicitly observed because the transition cross-section of hydrogen for the 1.468 eV photon excitation which is generally used is much smaller than that of silicon. Such a state has been observed by ultraviolet photoemission spectroscopy (UPS), and a number of sharp structures have been found in the region 5–10 eV from the top edge of the valence band [10]. It has also been reported that when the region of the valence-band edge is closely examined by UPS, the valence band edge is shifted toward the lower-energy side as the content of bonded hydrogen is increased [10]. This may be interpreted qualitatively as the partial replacement of Si–Si bonds with Si–H bonds of greater bonding energy.

Figure 4.3 shows the dependence of the band-edge energy on the content of bonded hydrogen, as calculated by the coherent potential approximation (CPA), which is form of mean-field approximation [11]. In contrast to the minor effect of hydrogen on the conduction-band-edge energy, it has been shown that the presence of hydrogen causes the valence-band edge to move towards the lower-energy side by removing the Si–Si bonding orbitals in the neighbourhood of the valence band edge. Consequently, the band gap is continuously expanded as the content of bonded hydrogen increases.

As mentioned in Chapter 3, the presence of bonded hydrogen is thought to make the network more flexible and to relax the structural disorder in the bond angle, bond length and dihedral angle, by reducing the average coordination number. On the other hand, the random replacement of Si–Si bonds with Si–H bonds may cause new structural disorders [12, 13]. In consideration of the inhomogeneous structures associated with bonded hydrogen, the effect of hydrogen on the electronic structure of a-Si:H is thought to be very complicated, and at present no satisfactory theoretical treatment has been established.

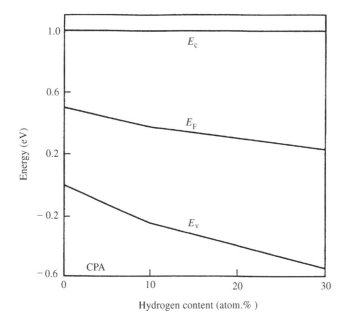

Fig. 4.3 The dependence of band-edge energy on hydrogen content as obtained from theoretical calculations [11, reproduced with permission. Copyright 1981 by the American Chemical Society]

4.1.3 ELECTRONIC STATE NEAR THE BAND EDGE

(a) Concept of the band tail

The effect of structural disorder is manifested most markedly in the vicinity of the band edge. The most important point is the formation of a localized band tail, and the presence of a mobility edge which separates it from the extended states. While the details of the band tail in a-Si:H will be described in connection with the electrical and optical experimental data in Sections 4.2 to 4.4, the concept of the band tail will be briefly illustrated here by using a simple model [15].

The band-edge energy is thought to undergo spatial fluctuation owing to the structural disorder. Let us regard this fluctuation as being an effective potential to which the band-edge electron is sensitive, constituted by an assembly of three-dimensional potential wells of spherical symmetry, as schematically shown in Fig. 4.4. We will try to see how the DOS is formed *en masse* by the bonded states created in quantum wells of random depth. The transfer of electrons between the adjacent wells (the tunnelling effect) is regarded as being negligible, i.e. the wave function of the electron is considered to be strongly localized in the vicinity of each well.

As given in a standard textbook of quantum physics, there is at least one bonded state if the potential V is greater (or deeper) than $V_C = \hbar^2\pi^2/8m^*a^2$, where $\hbar$ is the Planck constant, m^* the effective mass of the electron and a the radius of

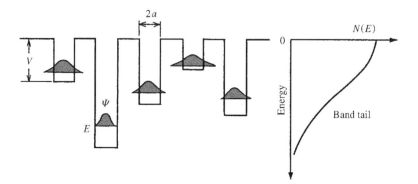

Fig. 4.4 A simplified model of potential fluctuations at the band edge and band tail states

the spherical potential well. The eigenenergy E of the ground bonded state (as measured in the direction of the energy axis in Fig. 4.4) is given approximately by the following expressions, depending upon the depth of the well V:

$$E \equiv \frac{m^*a^2}{2\hbar^2}(V - V_c)^2 \quad \text{(for a shallow well)} \tag{4.1}$$

$$E \equiv V - V_c/4 \quad \text{(for a deep well)} \tag{4.2}$$

$$E \equiv AV^2 - B \quad \text{(for a well of intermediate depth)} \tag{4.3}$$

where A is a constant proportional to $2m^*a^2/\hbar^2$. In expression (4.1), the localization length λ representing the extent of the wave function is almost proportional to $(V - V_c)^{-1/2}$. In the case of shallow wells, where λ is a few times larger than a, this discussion fails to be valid, since the adjacent wells cannot be regarded as being independent. It is known, under the conditions of expressions (4.2) and (4.3), that the largest value of λ is ca 1.5a.

If the probability distribution of the potential depth V is denoted by $p(V)$, the DOS formed by the above mentioned assembly of eigenvalues, $N(E)$ will be given by the following:

$$N(E) \approx p(V(E))\mathrm{d}V(E)/\mathrm{d}E \tag{4.4}$$

where $V(E)$ represents inverse functions for expression (4.1) and others. If $p(V)$ is assumed to be a Gauss function centered at $V = 0$, variance W^2, then the expressions (4.1)–(4.3) may be rewritten as follows:

$$N(E) \approx \exp\left(-2V_c^{3/2}\sqrt{E}/\pi^{1/2}W^2\right) \quad \text{(for a shallow region)} \tag{4.5}$$

$$N(E) \approx \exp[-E/2AW^2] \quad \text{(for an intermediate-depth region)} \tag{4.6}$$

$$N(E) \approx \exp[-E^2/2W^2] \quad \text{(for a deep region)} \tag{4.7}$$

Therefore, the shape of the DOS changes from an exponential form of $E^{1/2}$ to one of E^1 and finally to one of E^2, as the energy depth increases. According to a more detailed theoretical calculation, it has been confirmed that the region represented by equation (4.5), known as the Halperin–Lax region, is very narrow, while the represented by equation (4.6), a simple exponential function, is dominant over a few orders of magnitude of the DOS [17, 18]. This conclusion agrees well with the experimental data which will be described later. The DOS of Gaussian shape (equation (4.7)) is difficult to observe because it is actually covered by a contribution from the structural defects.

(b) Qualitative discussion on localization

The discussion so far has been made under the implicit understanding that the band tail is in the Anderson localized state [19]. However, if in Fig. 4.4 the adjacent wells are separated from each other by a distance which is as the expansion of the respective eigenwave function, then the energy states are linked up through the tunnelling effect. If this occurs successively for a full large of well arrays, the wave function of the electron becomes extended over the entire system. It is significant, therefore, to confirm that the band tail can be localized [20].

Let us denote the energy-level difference between a quantum well (to be designated as a site), and another site separated from the former by a distance r, as Δ and the strength of the coupling between these levels by $V(r)$. Therefore, an electron can move between the two sites if $V(r) > |\Delta|/2$. Assuming that $V(r) \sim V(0)\exp(-2r/\lambda)$, and the site energy is uniformly distributed over a width W following the ideas of Anderson, in place of the Gaussian distribution postulated in Section 4.1.3(a), the probability $P(r)$ of finding a site to which an electron can move for the first time, within a range of distance r from a site, is given by the following:

$$\frac{d}{dr}P(r) = \frac{2V(0)}{W}\rho\exp(-2r/\lambda)[1 - P(r)] \tag{4.8}$$

where ρ represents the number of sites in a unit volume. If equation (4.8) is integrated under the condition where $P(0) = 0$, then:

$$P(\infty) = 1 - \exp(-2\pi V(0)\lambda^3\rho/W) \tag{4.9}$$

From this expression, it is evident that $P(\infty)$ cannot be equal to 1, unless $W = 0$. In other words, if there is disorder, then the probability for an electron in a site failing to move to another site is finite, i.e. it can remain localized at that site. While the nature of the localized energy level will be described later in Section 4.3, it has been intuitively demonstrated that the vicinity of the band tail is localized, at least in the region which is deeper than the exponential DOS. Even in a system involving a disorder, there is an eigenstate and it is not conceivable that both localized and extended states co-exist at one and the same energy. Accordingly, there should be an energy position for separating them, which is called the mobility edge.

(c) The density-of-states in the vicinity of the band edge and mean free path

In this Section, the vicinity of the band edge in a system involving weak disorder will be examined by using a tight-binding Hamiltonian model based on the simple lattice. The electronic state of a system is assumed to be described by a linear combination of Wannier states $|i>$ localized at regularly arranged atomic or molecular orbital sites. If it is assumed that the site energy takes a random value v_i, and the transfer energy between the nearest neighbor sites is constant at V, then the Hamiltonian is given by the following:

$$H = V \sum_{i \neq j} |i><j| + \sum_i v_i |i><i| \tag{4.10}$$

As is well known, the first term in the above, the off-diagonal, gives the energy dispersion relationship of the regular lattice system $E = E_0(k)$, and the width of the created band, B, is represented by $2z|V|$, where z is the coordination number. Let us assume that the diagonal component v_i of the second term follows a Gaussian distribution with zero mean and variance W^2, and consider this as the perturbation term of disorder for the regular system represented by the first term in equation (4.10).

The DOS spectrum in the vicinity of the band edge is determined in the following way according to the renormalized secondary perturbation method [21]. It is assumed that the DOS of a non-perturbed system is proportional to $E^{1/2}$ near the band edge.

$$N(E) \approx \frac{V^{-3/2}a^{-3}}{2\sqrt{2}\pi^2} \left(\sqrt{(E - E_b)^2 + \sigma_i(E)^2} + E - E_b \right)^{1/2} \tag{4.11}$$

where a is the interatomic distance, E_b a negative value close to $-4W^2/B$ which corresponds to the shift in the DOS spectrum, and $\sigma_i(E)$ an imaginary self-energy component of, representing 'blur' in the spectrum. For $E > 0$, the blur is given by the following:

$$\sigma_i(E) = \frac{W^2}{4\pi} V^{-3/2} \sqrt{E} \tag{4.12}$$

From equation (4.11), it is evident that in the energy region where $E > E_b + cW^3/B^2$, where the constant c is close to 4 if the coordination number is 4, then the DOS resembles that of a non-perturbed system shifted by E_b, i.e. it behaves as the function $(E - E_b)^{1/2}$. While it is difficult to show explicitly by using such a simple analysis, it is expected that a band tail of the exponential form described in Section 4.1.3(a) follows the energy behavior mentioned in the above toward the lower-energy side [22].

The basic concept of the above discussion is that the electron state of a disordered system is composed of superimposed Bloch states of its 'regular' counterpart. The mean free path l constitutes an index for the proportion of included k-states, as defined by the following:

$$E - (E_b + i\sigma_i) = E_0 \left(k - \frac{i}{2l} \right) \qquad (4.13)$$

which may be approximated as:

$$l \approx 4\pi a (V/W)^2 \text{ if } E > E_b \qquad (4.14)$$

The range of k to be mixed, δk, is of the order of $1/l$, and its center of gravity, $\langle k \rangle$, is calculated by the following expression:

$$\langle k \rangle \approx \frac{1}{\sqrt{2Va}} \left[\sqrt{(E - E_b)^2 + \sigma_i(E)^2} + E - E_b \right]^{1/2} \qquad (4.15)$$

This equation gives an apparent dispersion relationship for a disordered system where $E = E(\langle k \rangle)$. In consideration of the effective mass m^* defined at the band edge of a non-perturbed system, which is given in the tight binding model of a simple lattice by $\hbar^2/2Va^2$, it seems that even in a disordered system the same effective mass may be assigned if $E > E_b + cW^3/B^2$.

In equation (4.6), if the effective mass m^* is replaced with a transfer energy V or bandwidth B, it is evident that the slope of the exponential tail E_0 is proportional to W^2/V or W^2/B. While it is by no means possible to describe the electron structure of actual a-Si:H with the simple model given above, this concept will be extremely useful in understanding the optical and electrical properties to be discussed in Sections 4.2 and 4.3.

4.1.4 LOCALIZED LEVELS BASED ON STRUCTURAL DEFECTS

(a) Structure of dangling bonds and thermal equilibrium statistics

The major structural defects in a-Si:H give an ESR signal of $g = 2.0055$, with these considered to be dangling bonds (to be abbreviated hereinafter as DBs). A number of different opinions concerning these were presented in Chapter 3. Dangling bonds involving unpaired electrons can capture either electrons or holes, thus being charged negatively or positively, respectively as directly proved on the basis of carrier transport characteristics [23]. In this way they can assume three states, which are usually designated as D^0 (neutral, filled with an electron), D^- (negatively charged, filled with two electrons) and D^+ (positively charged, empty state), respectively.

The wave function for the D^0 electron state, i.e. the unpaired electron, has been studied in detail by the use of ESR spectroscopy, and in particular, through the hyperfine interaction with ^{29}Si (nuclear spin, 1/2) [25, 26]. The results indicate that the wave function for a DB consists of an s-orbital component (ca 10%) and a p-orbital component (ca 90%), and is localized at the central Si atom at a probability of 50–80%. While this is supported by theoretical calculations [27], the electron nuclear double resonance (ENDOR) spectroscopic technique suggests that it is a more extended wave function [28]. It may be claimed without objection that the p-component is dominant in comparison to the pure sp^3 state (with an s-component of 25% and a p-component of 75%). For D^+ and D^-, neither any theoretical nor experimental data have been established up till now, although some researchers point out that D^+ and D^- are in the orbital states, sp^2 and p, respectively [29, 30] (see Chapter 3).

Let us denote the energy required for adding an electron to the D^+ state to produce the D^0 state by $E(+/0)$, and that required for adding another electron to the D^0 state to produce the D^- state by $E(0/-)$. Thus, the difference between the two energy levels is defined as the effective electronic correlation energy, U_{eff}, as follows:

$$U_{eff} = E(0/-) - E(+/0) \tag{4.16}$$

The correlation energy is a sum of the positive Coulomb energy U_c required for fitting two electrons into a localized DB orbital with a negative energy associated with lattice relaxations [31], electron–phonon interactions [32], and rehybridization of the DB orbitals [29, 30]. (see equation (5.11) in Section 5.3.3). On the basis of the single-electron scheme, the DB may be regarded as forming two levels at $E(+/0)$ and $E(0/-)$, which are designated as the D^+/D^0 and D^0/D^- levels, respectively.

The probability, or distribution function F^μ, of the DB being in the D^μ state ($\mu = +, 0, -$) under thermal equilibrium, is given by the following expressions according to simple statistical mechanics [33]:

$$F^+ = 1/Z \tag{4.17}$$

$$F^0 = 2\exp[-(E(+/0) - E_F)/k_B T]/Z \tag{4.18}$$

$$F^- = \exp[-(E(+/0) + E(0/-) - 2E_F)/k_B T]/Z \tag{4.19}$$

where

$$Z = 1 + 2\exp[-(E(+/0) - E_F)/k_B T]$$
$$+ \exp[-(E(+/0) + E(0/-) - 2E_F)/k_B T] \tag{4.20}$$

and E_F denotes the Fermi level. These distribution functions are illustrated in Fig. 4.5 for both positive and negative values of U_{eff}. It is evident that when $U_{eff} > 0$, D^0 is dominant if $|E(+/0) - E_F| < (U_{eff}/2)$. This means that if the Fermi level is within this range with respect to the D^+/D^0 level, then an unpaired electron can exist and DBs can be active in the ESR analysis. Moreover, for example, the Fermi level can be shifted very smoothly on doping within this energy range [34]. On the contrary, if $U_{eff} < 0$, the DBs are almost fully charged, and hence inactive in ESR, and the Fermi level is difficult to shift. In the case of a-Si:H, it is generally conceived that $U_{eff} > 0$, because the ESR signal can be observed under conditions of thermal equilibrium, its dependence on temperature is limited, and the signal intensity depends upon the level of doping [35].

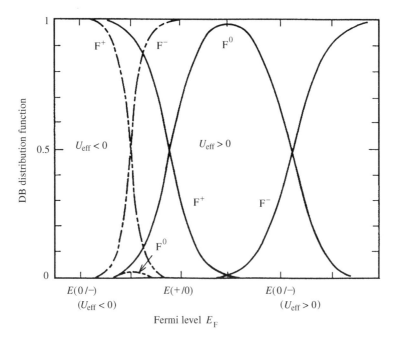

Fig. 4.5 Dependence of the electron-occupation factor (thermal equilibrium distribution function) of the dangling bonds on the Fermi level [33, reproduced with permission]

(b) Lattice relaxation and correlation energy

The defect-level energy is intensively affected by the local environment surrounding it. For example, in the case of an electron–phonon interaction based on a simple deformation potential model, the energy may take a form proportional to the generalized coordinate of the localized phonon, Q. Accordingly, the energy of a defect combined with a phonon $E(Q)$ may be written with relation to E_L, the defect energy without the combined phonon, i.e. the depth from the conduction band edge, as shown in Fig. 4.6(a), given as follows:

$$E(Q) \cong E_C - E_L + AQ^2 - BQ \tag{4.21}$$

where the term Q^2 corresponds to the bonded potential of the phonon. The coordinate Q at which $E(Q)$ is minimized, i.e. the defect is best stabilized, is equal to $B/2A$, and the defect level becomes deeper than E_L by $B^2/4A$ as viewed from the conduction band edge. Let us designate $B^2/2A$ as the lattice relaxation energy E_r. Since the optical electron transition between a defect and a conduction band occurs adiabatically (the Franck–Condon principle), the energies of light absorption and light emission become $(E_L + E_r)$ and E_L, respectively, as shown by the solid lines in Fig. 4.6(b). Therefore, the lattice relaxation energy corresponds to the well-known Stokes shift. If the thermal electron transition which takes place in the perturbation process involves the phonon kinetic energy term ignored in

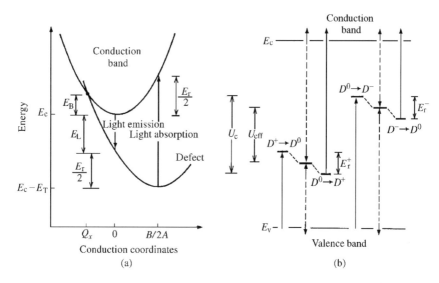

Fig. 4.6 (a) Coordination (coordinates) model of defect levels. (b) Various carrier transition processes occurring via dangling bonds: (————) optical process; (- - - - -) thermal process

equation (4.21) (the multiple phonon process), its transition energy is equal to $(E_L + E_r/2)$, as shown in Fig. 4.6(a).

In Fig. 4.6(b), the above discussion is applied to the DB system. In this figure, only part of the possible transitions are illustrated (in the form of arrows) to avoid complication. If lattice relaxation is considered, the correlation energy U_{eff} is approximated by the following:

$$U_{eff} \approx U_c - (E_r^- + E_r^+)/2 \tag{4.22}$$

These physical parameters have been determined in a-Si:H by the combination of sub-band optical absorption and various local-level spectroscopic studies [36–39], with the following values having been reported: $E_r^- \approx E_r^+ \approx 0.1$ eV, $U_c \approx 0.38$ eV, and $U_{eff} \approx 0.26$ eV. On the other hand, calculation of the total energy by applying the density-functional method to a 'super cell' containing DBs has yielded $E_r^- \approx E_r^+ \approx 0.53$ eV, and $U_c \approx 0.45$ eV and hence $U_{eff} \approx -0.2 \pm 0.2$ eV. In this way, the correlation energy is concluded to be zero or negative [31]. This calculation holds for a DB model placed in a particular environment, which not always represents DBs in actual a-Si:H. In fact, another calculation with spin polarization in different DB arrangements taken into consideration has reported $U_{eff} \approx 0.4 \pm 0.2$ eV [40].

If the correlation energy U_{eff} is negative or positive, with the Fermi level shifted to the vicinity of the band edge, charged DBs without unpaired spins will be dominant in the thermal equilibrium process, with little D^0 states and no ESR signal being observed. However, in the event of the transition $D^\pm$ (electron, or positive hole) $\longrightarrow D^0$, under illumination, an ESR signal becomes observable; this is known as light-induced ESR. In non-doped a-Si:H, signals of $g = 2.004$ and 2.01 are observed, which have been attributed to electrons or positive holes captured by the tails of the conduction and valence bands, respectively [41]. Recently, it has been pointed out that the signal involves a considerable proportion of DB-related components [42, 43]. Moreover, there has been a report of experiments which indicate that the signal is not related to the tail state, but to an electron state closer to a DB situation which causes an ESR signal in thermal equilibrium [44].

A detailed investigation combining depletion-layer-width-modulated ESR spectroscopy with the junction capacitance method [45] has yielded data suggesting that the U_{eff} of ordinary DBs in non-doped a-Si:H is almost equal to zero [46]. Possible interpretations based on these results include the following:

(1) A normal DB of $U_{eff} = 0$–0.3 eV co-exists with a negative DB, with the latter including charge-coupled $D^{\pm}$ intimate pairs [3].

(2) While U_{eff} is positive, a charged DB can exist because a-Si:H contains fluctuations in electrical potential attributable to structural heterogeneity, or because the DBs are distributed over a wide area in the gap (the so-called distributed defect pool) [48]. In the thermodynamic model of defects described in Chapter 3, it has been assumed that such DBs are formed in film deposition at different energy positions.

(3) $D^{\pm}$ can exist as compound defects of $U_{eff} < 0$ in which defects are charge-coupled with residual impurities in a-Si:H, such as nitrogen and oxygen [42, 49].

The following discussion is based on a standard model in which there is a single type of DB with $U_{eff} > 0$ at least in non-doped a-Si:H.

(c) Energy position

A direct method for determining the localized level involving DBs is to observe the thermal relaxation process of junction capacitance or current, either through deep-level transient spectroscopy (DLTS) [50], which evaluates the transient junction capacitance by temperature scanning, or through isothermal capacitance transient spectroscopy (ICTS) [51], which evaluates directly the time response under isothermal conditions. Fig. 4.7 shows typical distributions of the localized level density obtained by applying the ICTS method to n-type a-Si:H. It has been demonstrated by experiments in which the depletion-layer-width-modulated ESR signals are combined with DLTS measurements [45], that this level is equal to the D^- level of the thermal-equilibrium ESR center with $g = 2.0055$, i.e. of an ordinary DB. From this figure, it is evident that the D^- level (D^0/D^-) forms a Gaussian-type density distribution of the FWHM (0.1–0.2 eV) at a position which is 0.5–0.6 eV below the conduction-band edge. The D^0/D^- level is located at *ca* 0.2–0.4 eV (which is equal to the effective electron interaction energy U_{eff}) below this position, i.e. around the middle of the gap.

However, the situation is not that simple. In the ICTS method, where the process of thermal emission of electrons from a localized level is observed as a function of the time t, the time is 'converted' into the level depth $(E_C - E)$ under the condition that the thermal emissivity $e_n(E)$ meets the condition $e_n(E)t = 1$, as follows:

$$e_n(E) = \nu_n \exp\left[-(E_C - E)/k_B T\right] \tag{4.23}$$

$$E_C - E = k_B T \ln(\nu_n t) \tag{4.24}$$

It should be noted here that equivalent experimental data may yield different level depths depending upon what magnitude is assigned to the attempt-to-escape frequency, ν_n. In the example shown in Fig. 4.7, the energy axis is selected on the basis of $\nu_n \approx 10^9 \text{ s}^{-1}$ (dependent on energy), evaluated from an independent experiment [52]. On the other hand, some authors claim, on the basis of DLTS experimental data equivalent to this example, that the D^- level (D^0/D^-) in n-type a-Si:H is located at 0.8–0.9 eV below the conduction-band edge [50]. The dispute on this is still ongoing, since no decisive evidence is available [51, 53]. This issue will be dealt with in Sections 4.2 and 4.4 by consideration of other experimental data.

The discussion so far has been concerned with n-type a-Si:H. It seems difficult, however, to apply the transient capacitance technique to non-doped a-Si:H

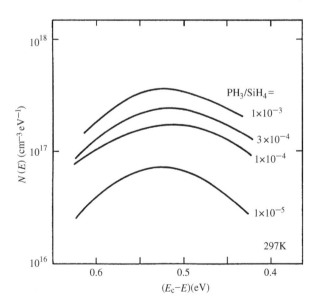

Fig. 4.7 Localized-states density distribution as determined by the ICTS method in n-type a-Si:H [51, reproduced with permission]

(to be designated as i-type hereinafter) because of longer dielectric relaxation times, or lower electric conductivities. For this reason, the transient current method or modulated photocurrent spectroscopy (MPCS) has been used for its evaluation [54–56]. The results obtained in this way indicate that the D^- level (D^0/D^-) is located at 0.5–0.6 eV below the conduction-band edge, as in the example shown in Fig. 4.7. It should be pointed out that the value of ν_n used in this calculation is of the order of 10^{12} s^{-1}.

In addition to the techniques mentioned here, the DB level has been estimated by using various other evaluation methods, details of which will be discussed in detail later; some results obtained from these are summarized in Fig. 4.8 in schematic form. In this figure, (a) shows i-type a-Si:H, thus providing a standard

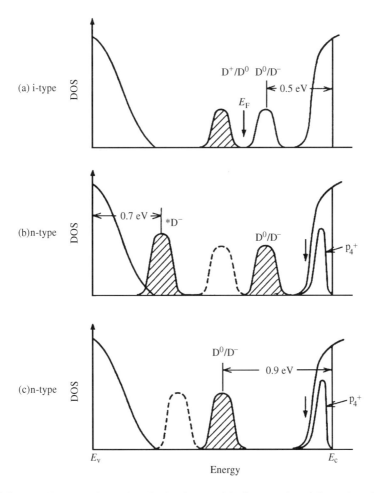

Fig. 4.8 Localized-level, density-distribution models for non-doped (i-type) and n-type a-Si:H

model for the ordinary DB, while (b) and (c) show n-type a-Si:H with $\nu_n \approx 10^9$ and $\approx 10^{13}$ s^{-1}, respectively. *D$^-$ in (b) represents a pair of charge-coupled defects involving a doped P-atom, P$_4{}^+$ − D$^-$, which is assumed to be formed in addition to the ordinary isolated D$^-$ at a position about 0.6−0.8 eV above the valence-band edge [57]. On the other hand, it can be interpreted from (c) that there is a single type of DB, and the D$^-$ level (D^0/D$^-$) is shifted by ca 0.2−0.3 eV downward from that of the i-type in the presence of P$_4{}^+$ [37, 58], or in response to thermodynamic requirements [59].

4.2 Structure of the Optical Absorption Edge

4.2.1 FEATURES OF THE OPTICAL ABSORPTION SPECTRUM

The absorption spectrum is usually determined from transmission and reflection spectra of thin films by considering multiple interference effects [60]. However, this technique is effective only in regions in which the condition $\alpha(E)d > 1$ holds, where $\alpha(E)$ is the absorption coefficient at light energy E and d is the film thickness, i.e. if $d = 10$ μm, then α is required to be $>10^3$ cm^{-1}. A number of techniques have been developed for determining the absorption spectrum in regions for which the absorption coefficient is smaller, or the light energy is lower. Representative methods include photoacoustic spectroscopy (PAS) [61], photothermal deflection spectroscopy (PDS) [62] measures the heat produced by the non-radiative recombination of light-induced carriers resulting from light absorption, the constant photocurrent method (CPM) [63], and modulated photo-current spectroscopy (MPCS) [64], which detects the photocurrent.

Figure 4.9 shows a typical example of the absorption spectrum of non-doped a-Si:H, obtained by combination of the light transmission spectrum with the photocurrent measurements mentioned above. In this figure, region A is gener-ally termed the 'Tauc region', where the spectral form can be approximated by equation (4.25) [65] and is attributed to the optical electron transition between the extended valence band and the conduction band:

$$\alpha(E)E \propto (E - E_0)^2 \tag{4.25}$$

where E_0 is termed the optical energy (or Tauc) gap, and is normally used for defining the energy gap of amorphous semiconductors. In some cases, it is used for measuring light energy in situations where the absorption coeffi-cient is $10^3 - 10^4$ cm^{-1}; sometimes, the second power on the right-handside of equation (4.25) is replaced by the third power. In subsequent discussions, the energy defined by equation (4.25) will be regarded as the optical gap. In Fig. 4.9, it can be seen that the position of the energy gap (E_0) is shifted toward the lower-energy side as the substrate temperature T_S is raised. It has been demonstrated both theoretically [11] and experimentally [66], that this shift can be closely correlated with a decrease in the bonded-hydrogen content of the film (and also with the density of the amorphous network).

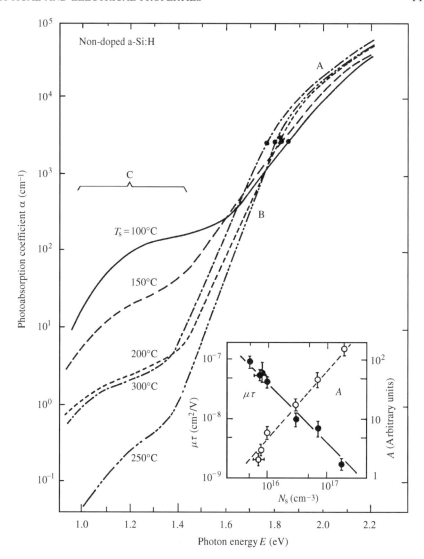

Fig. 4.9 Light absorption spectra of non-doped a-Si:H prepared at different substrate temperatures; inset shows the relationships of the absorption in region C (designated A) and the mobility–lifetime product, to the spin density [64, reproduced with permission]

Region B in Fig. 4.9 is known as the 'Urbach region' and has an exponential form given by the following:

$$\alpha(E) \propto \exp(E/E_u) \tag{4.26}$$

where E_u is called the Urbach tail slope. According to the discussion given in Section 4.1, this region can be described in terms of a transition between a band-tail state and an extended band. Furthermore, region C in Fig. 4.9 can be attributed to optical transitions involving structural defects (as shown in Fig. 4.6, for example). Descriptions of these three regions will be presented below, together with relevant theoretical considerations.

4.2.2 INTER-BAND TRANSITION REGION

(a) Non-direct optical transition model

The dielectric function ε_2 and the absorption coefficient α of the polarization vector e for light is given by equation (4.27). (In the case of non-polarized light, a direction averaging factor of 1/3 must be included.):

$$\varepsilon_2(E) = [c\hbar n(E)/E]\alpha(E) \tag{4.27}$$

$$= (2\pi e)^2 \left(\frac{\hbar}{m_0 E}\right)^2 \frac{2}{\Omega} \sum_{i,f} |\langle i|pe|f\rangle|^2 \delta(E_f - E_i - E) \tag{4.28}$$

$$= (2\pi e)^2 \frac{2}{\Omega} \sum_{i,f} |\langle i|re|f\rangle|^2 \delta(E_f - E_i - E) \tag{4.29}$$

where $n(E)$ is a real component of the refractive index, and $|\mu>$ is a legitimate eigenstate of eigenenergy E_μ normalized by the volume Ω, with $\mu = i$ representing the valence band and $\mu = f$ the conduction band. Expressions (4.28) and (4.29) represent the momentum matrix and dipole matrix elements, respectively, of the dipole-type transition approximation. With these denoted by M_{if}, the following function can be defined:

$$I(E) = \frac{2}{\Omega} \sum_{i,f} M_{if}^2 \delta(E_f - E_i - E) \tag{4.30}$$

If an averaged transition matrix is introduced, then we obtain:

$$\overline{M_{CV}^2}(E) = \sum_{i,f} M_{if}^2 \delta(E_f - E_i - E) \bigg/ \sum_{i,f} \delta(E_f - E_i - E) \tag{4.31}$$

Equation (4.30) can then be rewritten as follows:

$$I(E) = \frac{2}{\Omega} \overline{M_{CV}^2}(E) \sum_{i,f} \delta(E_f - E_i - E)$$

$$= \frac{\Omega}{2} \overline{M_{CV}^2}(E) \int N_V(E') N_C(E' + E) dE' \tag{4.32}$$

where $N_\mu(E')$ is a legitimate electron-state density function. This expression applies exactly to both the crystalline and amorphous states. For the crystalline state, where the wave vector k represents a suitable quantum number corresponding to the momentum, this relationship must be conserved in the allowed direct-type optical transition (without any phonon intervention). For this reason, $\overline{M_{\text{CV}}^2}(E)$ has a strong dependence on the light energy in the vicinity of the absorption edge which relates to $(E - E_0)^{-3/2}$.

In the case of the amorphous state, if its electron state is to be described by a superposition of the Bloch states for the corresponding crystal (unitary transformation), then the $\boldsymbol{k}$ states within a range of $1/l$ (where l is the mean free path) will be involved, as stated in Section 4.1. It can be assumed that the optical transition processes in the amorphous semiconductor are as shown schematically in Fig. 4.10. Assuming that the mean free path is around a mean interatomic distance (2.3–2.4 Å in a-Si:H), the optical transition may occur uniformly at any value of $\boldsymbol{k}$, but only if the energy is conserved. This is called a non-direct optical transition. In the amorphous semiconductor, it is often claimed that the selection rule for $\boldsymbol{k}$ is relaxed, or that the rule of conservation of momentum does not hold. This is an apparent situation in cases where the amorphous state is likened to the crystalline state.

Under these extreme conditions, a random-phase approximation is applicable, in which the phase of the wave function is randomly variable for each of the atom positions, and consequently, it can be demonstrated that $\overline{M_{\text{CV}}^2}(E)$ takes a

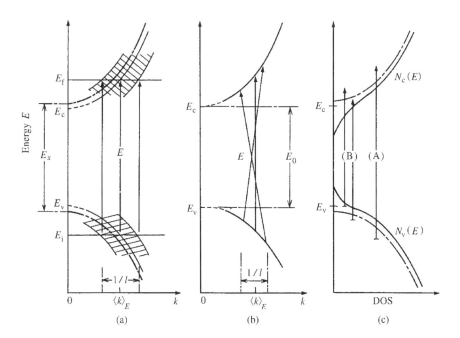

Fig. 4.10 Models illustrating the optical transition processes in amorphous semiconductors: (a) model which takes consideration of 'blurring' in E–k dispersion relationship due to the presence of disorder (see Section 4.1.4); (b) indirect transition model; (c) DOS model for the transition process in the optical absorption areas, A and B

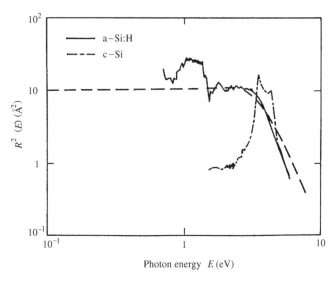

Photon energy E (eV)

Fig. 4.11 Light-energy dependence of the normalized mean dipole transition matrix R^2 in a-Si:H and c-Si; R^2 is defined as $R^2_{CV}N/2$ [4, reproduced with permission. Copyright 1985 by the American Physical Society]

certain value, i.e. $zM^2/2N$, where z is the coordination number, and N is equal to the number of atoms in a volume Ω [67]. Here, the dipole matrix element in equation (4.29) is used as the transition matrix, because optical transitions occur owing to the superposition of the valence band and the conduction band at the same atom site. It has been demonstrated that the dipole matrix element has no appreciable energy dependence, at least in the Tauc region, by the experimental results shown in Fig. 4.11, which are obtained from the results presented in Fig. 4.2 and the determination of the dielectric function ε_2 [4], discussions based on the summation rules for transition matrix elements [68], and detailed theoretical calculations [69].

(b) Optical energy gap

If a parabolic DOS is postulated for the energy region remote from the band edge, as stated in Section 4.1, then equation (4.32) can be simplified to give the following, which derives from equation (4.28):

$$\varepsilon_2(E) \propto (E - E_0)^2 \qquad (4.33)$$

which is distinctly different from equation (4.25), because in the case of the latter equation (4.28) is applied on the assumption that the averaged momentum transition matrix element is constant. Although it has been recognized that equation (4.33) has a more precise physical meaning and is a better approximation (as shown in Fig. 4.12), in the sense that it can represent broader spectral regions [4, 68–70], and the gap E_0 derived from the two expressions has different values,

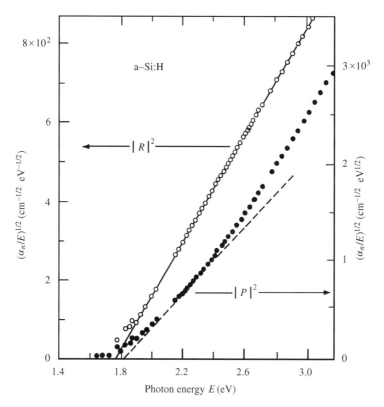

Fig. 4.12 Examples of absorption-coefficient spectral plotting corresponding to equations (4.33) and (4.25), assuming that the dipole transition matrix R^2 and the momentum transition matrix P^2 are constant [70, reproduced with permission]

at the present time it is more customary to use equation (4.25) because of the easier evaluation.

This argument may support the facts that the mean free path l is naturally close to the mean interatomic distance a, and that the k-selection rule is totally abolished in the case of amorphous semiconductors. It has been pointed out, however, on the basis of theoretical considerations, that in a-Si:H where the structural disorder is not so extensive, some traces of the rule may persist [69, 71–73].

Nevertheless, it has also been demonstrated that equation (4.33) holds approximately over a limited spectral region. This indicates that the mean free path is not always of the order of the interatomic distance [73, 74], which provides an important clue for understanding the transport phenomena (to be described later).

Let us now provide a simple expression for the optical gap. As stated in Section 4.1, structural disorder causes the band edge to shift by a magnitude proportional to the parameter of disorder W^2 (in addition to the blur in the DOS spectrum). It is supposed, therefore, that the gap E_0 can be approximated by the following [71, 73, 74]:

$$E_0 \approx E_X - \sum_{\mu=c,v} c_{0\mu} W_\mu^2 \tag{4.34}$$

where E_x represents the band gap of a non-perturbed system (the corresponding crystal), and c_0 is a constant which is weakly dependent on W. Since this relationship is based on a temperature of zero, it is necessary, in the case of finite temperatures, to consider the additional thermal disorder due to electron–phonon interactions. Assuming a distortion potential interaction, the disorder W_T^2 can be approximated by the following:

$$W_{T\mu}^2 \cong \frac{\gamma\mu}{2}\theta \coth\left(\frac{\theta}{2T}\right) \tag{4.35}$$

where γ is a quantity related to the magnitude of the deformation potential, and θ is the temperature converted from the effective Einstein frequency (corresponding to the center-of-gravity frequency of the longitudinal acoustic (LA) phonon spectrum). Equation (4.34) can be expanded by using the following relationship.

$$E_0(T) \approx E_X - \sum_{\mu=c,v} c_{0\mu}(W_\mu^2 + W_{T\mu}^2) \tag{4.36}$$

It is evident from the above equation that in the higher-temperature region the gap is reduced in proportion to the temperature T. It has been confirmed experimentally that the gap does have a temperature dependence in accordance with equation (4.36) over a wide temperature range [61, 75].

4.2.3 BAND-TAIL TRANSITION

(a) Urbach tail energy

It is believed that the region B in Fig. 4.9, where the absorption coefficient spectrum takes the shape of an exponential function, can be attributed to the optical transition which involves a band-tail state. As stated in Section 4.1, the band tail has an exponential form and its slope $E_{u\mu}$ is proportional to the parameter of disorder W^2. This may be approximated, therefore, as with equation (4.36), as follows:

$$E_{u\mu}(T) \approx c_{u\mu}(W_\mu^2 + W_{T\mu}^2) \tag{4.37}$$

If it is simply assumed that the optical transition spectrum between the band tail and the extended band state (with an almost parabolic DOS) is determined by the convolution integral for the two states, it appears that the DOS of the exponential tail, which is strongly dependent on the energy, is directly reflected in the spectrum, and if $E_{u\mu}$ is greater, then the spectrum corresponds to the Urbach tail E_u given by equation (4.26). This is not self-evident, but it should not be regarded as being a simple function of $E_{u\mu}$ for either or both of the valence and conduction bands [73, 76].

In cases where the disorder parameter for the valence band is assumed to be equal to that of the conduction band, the Urbach tail energy E_u can be represented by the following equation [68, 71, 73, 77]:

$$E_u(T) \approx c_u(W^2 + W_T^2) \tag{4.38}$$

where the factor c_u is given by $2c_{uc}c_{uv}/(c_{uc} + c_{uv})$. Fig. 4.13 presents examples of the temperature dependence of the Urbach tail energy E_u and the optical energy gap E_o [61]. This dependence agrees with that suggested by equation (4.38).

By combining equations (4.38) and (4.36), the following relationship is obtained:

$$E_o(T) = E_X - cE_u(T) \tag{4.39}$$

It has been demonstrated experimentally that equation (4.39) holds approximately for a-Si:H with $E_x \sim 2$ eV and $c \sim 6.2$ [75]. The value for factor c has also been confirmed by theoretical calculations [73, 74]. It should be noted, however, that the relationship (4.39) corresponds to a dependence on the temperature of measurement, and is not universally applicable to all samples of a-Si:H, as a consequence of the fact that E_x is sensitive to the amount of bonded hydrogen in the material.

Since the usual optical measurements described above consider the optical transition from the valence band to the conduction band, it is evident that the

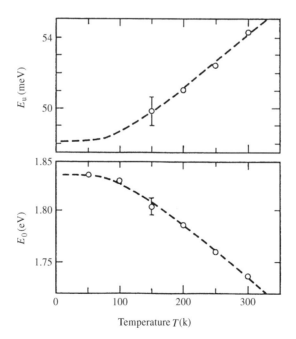

Fig. 4.13 Plots showing the temperature dependence of the Urbach tail energy E_u and the optical energy gap E_o in a-Si:H [61, reproduced with permission]

results do not reflect the actual DOS involving the spectrum. On the other hand, the photoelectric total yield based on monochromated ultraviolet irradiation represent the vacuum levels in which the final state of the optical transition is rather structureless, and it is possible to estimate directly the DOS for the electron-filled initial state [76]. Fig. 4.14 shows the dependence of the tail energies of the valence band (E_{uv}) and the conduction band (E_{uc}) on the measurement temperature, as determined by the photoelectric technique [77]. For the conduction band tail, the measured spectrum has been corrected by the use of the Fermi distribution function. It can be claimed from Fig. 4.14, in terms of equation (4.37), that the conduction band tail is governed by thermal disorder, while the valence band tail is governed by the original structural disorder within this temperature range. By comparing the temperature dependence of the Urbach tail energy E_u shown in Fig. 4.13 with that presented in Fig. 4.14, it is evident that E_u is not always uniquely determined by either of the band tails [73]. This point should be kept in mind when interpreting the optically measured data.

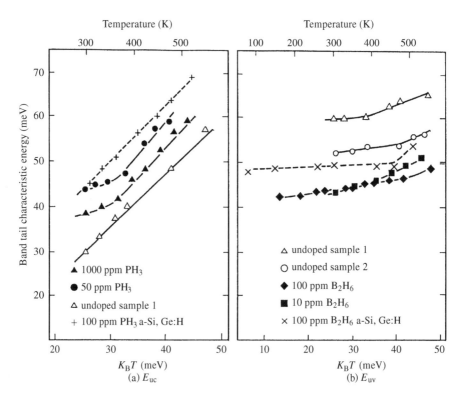

Fig. 4.14 Dependence of the tail energy of the valence band (E_{uv}) and conduction band (E_{uc}) on the temperature of measurement, as determined by the photoelectric total yield technique [77, reproduced with permission. Copyright 1990 by the American Chemical Society]

(b) Field-modulated spectra

When an electric field is applied to a semiconductor, the electron state is modulated and its effect is manifested in the vicinity of a critical point in the optical spectrum. In the case of a crystalline semiconductor, various types of information on the band structure can be gained from the modulation spectrum [78]. In a-Si:H, similar experiments have been conducted to determine $\Delta\alpha$ (abbreviated as the electro absorption, EA), and a broad single peak structure ($\Delta\alpha > 0$) has been reported in the vicinity of the optical energy gap, i.e. in the Urbach tail region of the absorption spectrum [79]. Recently, some experimental data of great significance (shown in Fig. 4.15) have been reported [80]. From these measurements, a distinct anisotropy is recognized in the signal intensity $\Delta\alpha$, depending upon whether the direction of the polarized light is perpendicular to or parallel to the direction of the applied field. Such an anisotropy is a characteristic of the amorphous semiconductor and is not observable in the crystalline semiconductor.

According to theoretical analysis in which the transition matrix elements are regarded as being not constant, when the polarization is parallel to the applied field, the transition matrix elements increase almost in proportion to the disorder parameter $W_c^2 W_v^2$ [73]. The additional $\Delta\alpha$ component resulting from this is summed with the isotropic $\Delta\alpha$ component derived from the mixing of the wavefunctions caused by the electric field, to make up the entire EA spectrum. This

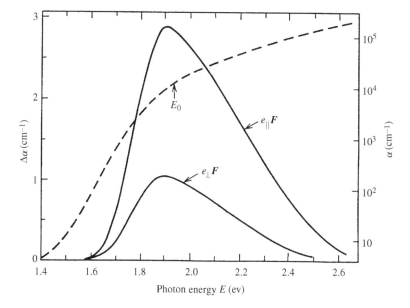

Fig. 4.15 An example of measurement of the polarization-dependent, field-modulated absorption spectrum; e represents the direction of light polarization and F the field vector [80, reproduced with permission]

means that the anisotropy of the EA spectral intensity is intensively manifested in the amorphous state only, thus providing a measure for evaluating the disorder.

The theoretical calculations indicate that the degree of anisotropy of the EA signal intensity is given as shown in equation (4.40), using the values of the mean free path in the valence and conduction bands:

$$\Delta\alpha_{\parallel}/\Delta\alpha_{\perp} \approx 1 + \frac{98}{3}\pi^2 \left(\sqrt{l_c/a} + \sqrt{l_v/a}\right)^{14} \qquad (4.40)$$

When this relationship is applied to the experimental data shown in Fig. 4.14 and the averaged dipole matrix elements given in Fig. 4.11 are treated in a similar way, the mean free paths are estimated to be $l_c \sim 17$ Å, and $l_v \sim 4$ Å, and the effective masses of the unperturbed reference crystal system as $m_c^* \sim 0.3m_o$ and $m_v^* \sim 0.7m_o$ [73]. The validity of these estimations will be examined later.

4.2.4 OPTICAL TRANSITIONS CAUSED BY STRUCTURAL DEFECTS

(a) Absorption intensity and dangling-bond density

The absorption band observed at around 1.2 eV (light energy) in Fig. 4.9 is related to the structural defects which are located deep in the forbidden band. The integrated intensity of the additional absorption in non-doped a-Si:H (usually denoted by A) [81] is compared with the spin density of $g = 2.0055$ (see the inset in Fig. 4.9) in Fig. 4.16 [61]. It is evident that the two parameters are distinctly

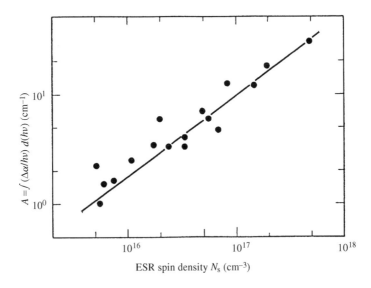

Fig. 4.16 Relationship of the integrated absorption intensity A in the optical absorption region C to the ESR spin density N_S (involving dangling bonds) [61, reproduced with permission]

proportional. It can therefore be said that within the range of the standard model described in Section 4.1, this absorption band is based on the neutral-dangling bond state D^0, filled with a single electron. The optical transition corresponds to $D^0 \longrightarrow D^+ + e(D^0-CB$ transition) and $D^0 \longrightarrow D^- + hole(h)$ (VB–D^0 transition). From the linear relationship shown in Fig. 4.16, the factor of proportionality (roughly speaking, the mean value of the optical cross-section) is found to be of the order of 10^{-16} cm^2. This is supported by results obtained from transient optical capacitance measurements [82] or photo-induced absorption spectroscopy [83], combined with ESR spectroscopy measurements, which show that the additional absorption band originates from an optical transition.

(b) Dangling-bond levels in i-type a-Si:H

The DB levels can be estimated from absorption bands which involve DB defects. In this case, the level locations measured from the band edge can be defined directly by the light energy. It should be noted, however, that the energy determined as shown in Fig. 4.6, is different from that obtained through thermal process measurements (DLTS and ICTS) by the energy related to lattice relaxation.

The optical cross-section for a single level $\sigma_{opt}(E)$, i.e. the spectral shape of the absorption band, is given by the following:

$$\sigma_{opt}(E) \propto (E - E_0)^\alpha / E^\beta \tag{4.41}$$

where E_0 is the threshold of the optical transition, which corresponds to the optical level depth; α and β are constants which relate to the electron state and charge conditions of the defect, respectively, which are estimated for the transition between the neutral defect and parabolic-type band, i.e. $\alpha = 1/2$, $\beta = 1$ [84], $\alpha = 3/2$, $\beta = 3$ (forbidden transition model) [85], or $\alpha = 1/2$, $\beta = 3$ (allowed transition model) [86]. It should be noted that the distribution of the defect states needs to be taken into consideration.

Figure 4.17 shows optical transition spectra which have been phase-separated by modulated photocurrent spectroscopy [87]. Each of the separated transitions have been identified as shown in the inset. Arrows in the figure indicate the thresholds of transition determined in accordance with the allowed transition model, at ca 0.9 eV for $D^0 \longrightarrow D^- + e(D^0-CB$ transition) and ca 1.2 eV for $D^0 \longrightarrow D^- + h$ (VB–D^0 transition). If the optical gap is set at 1.76 eV, the location of the conduction-band edge is estimated to be 0.6 eV for the D^0/D^- level and 0.9 eV for the D^-/D^0 level. Two other optical transitions involving dangling bonds, namely $D^- \longrightarrow D^0 + e$ (D^-–CB transition) and $D^+ \longrightarrow D^0 + h$ (VB–D^+ transition) have been determined by transient photocurrent measurements [39] and after applying the correction for lattice relaxation energy, the conclusions reached in Section 4.1. have been eventually deviced.

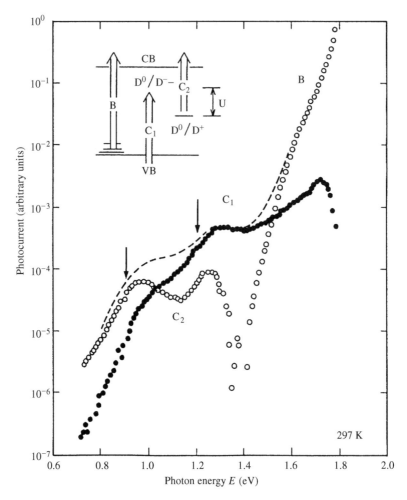

Fig. 4.17 Modulated photocurrent spectra of non-doped a-Si:H in the lower-light-energy region; spectra are separated on the basis of the differences in phase characteristics of the photocurrents corresponding to the optical transitions shown in the inset [87, reproduced with permission]

(c) Dangling-bond levels in n-type a-Si:H

DB transition spectra determined by the transient capacitance method are shown in Fig. 4.18, together with other relevant information [88, 89]. In this case, as the DBs are mostly in the D^- state, the only possible optical transition is $D^- \longrightarrow D^0 - e(D^- - CB)$. This figure shows two features in the transient capacitance response, namely one at ca 0.6 eV, corresponding to the D^0/D^- level in Fig. 4.7, and an other at ca 1 eV, which corresponds to the defect-absorption band of n-type a-Si:H and is regarded to be a charge-coupled defect, a $P_4^+ - D^-$ pair,

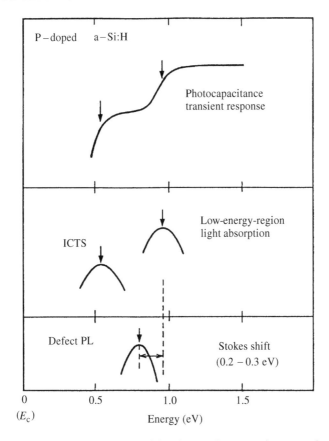

Fig. 4.18 DB transition spectra obtained by the transient capacitance method and other data concerning the localized levels in n-type a-Si:H; the energy corresponds to the depth from the conduction-band edge. [88, reproduced with permission]

corresponding to a *D$^-$ level. These data provide the experimental grounds for the model presented earlier in Fig. 4.8(b).

The researchers who presented the alternative model shown earlier in Fig. 4.8(c) have also reported results obtained from transient capacitance measurement [38], but their spectra show only single structures at *ca* 1 eV (see Fig. 4.18). Measurements of the defect absorption band by using the constant photocurrent method (CPM) [58] and optical modulation spectroscopy (OMS) [37] have yielded data supporting these experiments. The interpretation of the shift of the D^0/D$^-$ level by *ca* 0.2–0.3 eV, in comparison to i-type a-Si:H, has been described earlier in Section 4.1. The validity of the models (b) and (c) shown in Fig. 4.8 has not yet been settled, and an overall re-examination is still request for the method of measurement, the conditions and the analytical procedure.

4.3 Electrical Properties

4.3.1 MECHANISM OF ELECTRICAL CONDUCTION

(a) Band conduction and mobility edge

Discussions of this particular issue are usually presented in accordance with scaling or renormalization group treatments [90], and a number of excellent reviews on these approaches are available [20, 91]. Let us start here from what is known as the Kawabata expression for the conductivity [20, 92], which is derived from a microscopic view based on Kubo's theory:

$$\sigma(E) \cong \sigma_B + \Delta\sigma(E) \tag{4.42}$$

This expression gives the conductivity when the Fermi level is at energy E. The first term represents the classical Boltzmann term, which is given by equation (4.43) which uses the mean free path l and wavenumber k_E,

$$\sigma_B = \frac{e^2}{\hbar}\frac{1}{3\pi^2}lk_E^2 \tag{4.43}$$

The second term is a correction factor based on the multiple scattering effect resulting from the random scattering potential, and it is known to be as given below within the range of correction for the lowest order [20]:

$$\Delta\sigma(E) \cong -\frac{2e^2}{\pi\hbar}\frac{1}{\Omega}\sum_q \frac{1}{q^2} \tag{4.44}$$

The overall electrical conductivity can be expressed by equation (4.45) when the summation with respect to the wavenumber q is replaced by an integral with an upper limit of $1/l$ and a lower limit of $1/L$ (corresponding to the length of one side of the sample):

$$\sigma_L(E) \cong \frac{1}{\pi^3}\frac{e^2}{\hbar}\left[\frac{1}{\xi(E)} + \frac{1}{L}\right] \tag{4.45}$$

It should be noted that the coefficient $1/\pi^3$ may involve an error factor of $2-3$ [93]; ξ is called the correlation length and in this case is given by the following:

$$\xi(E) \equiv l\left[\frac{\pi}{3}(k_El)^2 - 1\right]^{-1} \tag{4.46}$$

The electrical conductivity observed for an actual sample (with the sample size L infinitely large in comparison to the mean free path) is expressed as shown below by using the Fermi distribution function $f(E)$:

$$\sigma(T) \cong \int \sigma_\infty(E)\left[-\frac{\partial f(E)}{\partial E}\right]dE \tag{4.47}$$

As is well known, the term in parentheses can be approximated by the Boltzmann term $(1/K_BT)\exp(-(E - E_F)/K_BT)$, in the case of non-degenerate semiconductors. Accordingly, the electrical conductivity resulting from electron transport in amorphous semiconductors can be formally described by the following:

$$\sigma(T) \cong \sigma_0 \exp[-(E_c - E_F)/K_B T] \tag{4.48}$$

where σ_0 is a pre-exponential factor for the electrical conductivity, and E_c represents the mobility edge (to be discussed below).

Now, let us examine equation (4.45) in detail. If the sample size L is comparable to the mean free path, the correction term (equation (4.44)) disappears, leaving only the classical Boltzmann term. It can be clearly seen that as the sample size grows, then the conductivity decreases in inverse proportion to L, and when it exceeds the correlation length approaches a fixed value in inverse proportion to this length. The problem here is the magnitude of this value. As expected from equation (4.47), this parameters governs the electrical conductivity of an actual sample. It is readily understood that at an energy value where the correlation length is infinitely large, the conductivity becomes zero. Therefore, equation (4.46) suggests that the current flows only in energy states which meet the following requirement:

$$K_E l \geq \sqrt{3/\pi} \tag{4.49}$$

The energy position at the boundary is the mobility edge E_c. Energy states higher than this value have extended wave functions, and those states which are lower are localized states which are characterized by finite localization lengths. It can be stated that the localization of an electron is the result of the quantum mechanical interference effects of the electron wave which are induced by random multiple scattering. (The Kawabata correction term is based on this concept.) Even in the extended state, the amplitude of the wave function appears to be fluctuating in the vicinity of the mobility edge. The scale of this fluctuation constitutes the correlation length, and when the sample size exceeds this range, the nature of the wave function, extended or localized, can be decided. In this region, the ohmic conductivity, which is not dependent upon the sample size, can be defined.

The term $k_E l$ in equations (4.46) and (4.49) can be evaluated by using the simple models discussed in Section 4.1, i.e. equations (4.14) and (4.15), thus yielding the following results [94]:

$$E_c \approx E_b + 4.5 \times 10^{-3} \frac{W^4}{|V|^3} \tag{4.50}$$

$$\sigma_\infty(E) \approx \frac{4}{3\pi} \frac{e^2}{\hbar a} \frac{|V|}{W^2}(E - E_c) \tag{4.51}$$

First, we shall attempt to apply equation (4.50) to a case where the disorder is very extensive, although this approach seems to be controversial, because the relationship has been derived from an assumption of weak disorder. When the mobility edge approaches the middle of the band, the band as a whole is localized. The condition for this situation is $E_c = B/2$. If it is assumed that $B = 2z|V|$ and $z = 6$ (a simple cubic lattice), and the variance W^2 of the Gaussian-type disorder is replaced by the variance $W_a^2/12$ of the Anderson box-type disorder, the Anderson localization condition is obtained as $W_a/B = 1.7$. This value agrees with that obtained by computer simulation calculations [95]. It should be noted that the discussion given here is based on a rough approximation which is valid only under certain conditions, and is applicable for surveying the qualitative trends and not to quantitative estimations.

Equation (4.51) indicates that the conductivity is zero at the mobility edge and continuously grows from this point, disregarding the coefficient and details of the energy dependence. Therefore, within the range of the zero-temperature theory where the electron–phonon interaction is ignored at least, Mott's well-known minimum metallic conductivity concept does not apply. By using equation (4.51), the value of the coefficient σ_0 in equation (4.48) can be readily calculated, and the average mobility of the band conduction μ_c can be obtained from $\sigma_0/ek_BTN(E_c)$. The mobility values for the electron and the hole, calculated by using the parameters estimated for a-Si:H in Section 4.2., (such as the mean free path) are ca 17 and 0.6 cm^2 V^{-1}s^{-1}, respectively [73, 74].

Finally, let us examine the conductivity at finite temperatures. There are two ways to deal with this situation, namely to explicitly consider the electron–phonon interaction, and to take in, conceptually the effect of the inelastic scattering which distorts the coherence of the electron wave. We shall discuss the latter method first. Inelastic scattering reduces the interference effect of the electron wave, which underlies the localization, together with the multiple scattering. The sample size L in equation (4.45) is to be replaced, therefore, by the inelastic scattering length L_i [94, 97]. If the latter length is shorter than the correlation length mentioned above, the conductivity assumes a constant value in the vicinity of the mobility edge. This situation is illustrated schematically in Fig. 4.19.

The conductivity which will be observed is given, using equation (4.47) by the following:

$$\sigma(T) \cong \frac{1}{\pi^3} \frac{e^2}{\hbar L_i} \exp[-(E_c - E_F)/K_BT] \tag{4.52}$$

where the pre-factor σ_0 is around $800/L_i(\text{Å})$, which includes an error corresponding to the uncertainty of the coefficient. The mean mobility μ_c is related to the prefactor σ_0 by $\sigma_0/ek_BTN(E_c)$.

According to a theory which renormalizes the electron–phonon interaction into a Hamiltonian for the single electron, which takes the form of a static coupling between them [98, 99], it has been shown that the conductivity given by equation (4.52) (with L replaced by the mean free path, as inelastic scattering is not taken into consideration) needs the addition of a term representing hopping conduction (to be described later). In the energy state below the mobility edge, hopping conduction alone can occur, i.e. at finite temperatures,

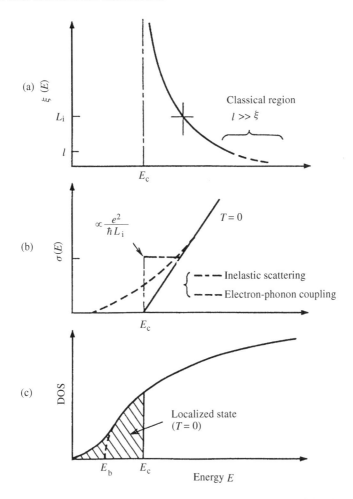

Fig. 4.19 Schematic representations of (a) the correlation length ξ of the electron wave function, (b) the differential conductivity $\sigma(E)$ and (c) the density-of-states distribution in the vicinity of the band edge, as a function of the energy E.

the mobility edge cannot be clearly defined owing to the electron–phonon interaction, which leaves a distinct finite bluring. This phenomenon is known as phonon-induced delocalization. Next, the energy position which contributes most to the carrier transport varies depending upon the temperature. It is generally accepted for electron conduction in a-Si:H, that the energy position which provides the major transport energy is located in the vicinity of the mobility edge (at temperature 0°C) at around room temperature, and is shifted toward the lower-temperature side as the temperature is increased [99]. For the sake of simplicity, therefore, E_c in equation (4.52) is to be regarded as representing not the mobility edge itself, but the temperature-dependent transport energy.

(b) Hopping conduction

As expected from the presence of the Boltzmann factor in equation (4.52), conduction in the extended state is reduced as the temperature is lowered, and in its place, hopping conduction, through the localized level close to the Fermi level, dominates. In the case of electron conduction in a-Si:H, the switching occurs at temperatures lower than *ca* 150 K [100]. In this case, the conductivity takes a form which is proportional to the hopping rate [101, 102]:

$$\Gamma = \nu_0 \exp(-2R/\lambda - W/k_B T) \qquad (4.53)$$

where ν_0 is around the phonon frequency level, λ is the localization length of the wave function for the localization level, and R and W represent the spatial and energy intervals, respectively, between the hopping-pair levels. The mobility is calculated by multiplying this expression by the term $eR^2/6k_B T$. This holds only for a specific hopping-site pair, and in order to discuss the conductivity and mobility that are actually observed, a statistical treatment is required which considers the Fermi distribution, density-of-states distribution, energy dependence of the localization length and details of the conduction channel.

If the conduction occurs in the vicinity of the Fermi level (state density, N_F) and is governed by hopping between a pair of levels, defined by R or W, which is determined by minimizing the exponent of equation (4.53) under the condition of $4\pi R^3 W N_F/3 = 1$, the following relationship is obtained:

$$\sigma \propto \exp[-(T_0/T)^{1/4}] \qquad (4.54)$$

where T_0 is a function of λ and N_F. This expression is known as the Mott equation of variable range hopping [102], and has been demonstrated experimentally in both non-hydrogenated a-Si and ordinary a-Si:H at very low temperatures [103]. It has been shown theoretically that such a temperature dependence does not occur for hopping conduction between localized states which have a negative effective correlation energy [104], and it may be claimed, therefore, from this point of view, that for localized states around the Fermi level in a-Si:H, the condition $U_{eff} < 0$ does not apply.

4.3.2 DC CONDUCTIVITY AND THERMOELECTRIC POWER

(a) DC conductivity

The direct current conductivity $\sigma(T)$ in a-Si:H has, in general, a temperature dependence of the activation type at around room temperature, based on transport in the vicinity of the mobility edge, as given in the following:

$$\sigma(T) \approx \sigma_0^* \exp(-\Delta E_\sigma^*/k_B T) \qquad (4.55)$$

where σ_0^* and ΔE_0^* are the pre-factor and activation energy to be measured, respectively. For these two parameters, the following approximations hold, in relation to equation (4.48):

$$\sigma_0^* \approx \sigma_0 \exp[(\gamma_F + \gamma_G + \gamma_T)/k_B] \tag{4.56}$$

and

$$\Delta E_\sigma^* \approx (E_c - E_F)_{T \to 0} \tag{4.57}$$

The three parameters in the exponent of equation (4.56) represent factors of proportionality for an assumed Fermi level (γ_F), the relative energy interval between the transport level and the Fermi level (γ_G), and the transport energy (γ_T), as linear functions of the temperature, respectively. First, the temperature dependence of the Fermi level is called the statistical shift, and always exists unless the DOS is symmetric across this level, or unless the DOS is fixed at localized levels of high density [105]. The second term, γ_G, is regarded as being equal to the temperature dependence of the forbidden band (the Tauc gap, for example), i.e. at about 5 k_B [106], multiplied by $\Delta E_\sigma^*/E_0$. Finally, the temperature coefficient of the transport energy, γ_T, is determined by the competition between the phonon-induced delocalization ($\gamma_T > 0$) mentioned above and the Boltzmann factor ($\gamma_T < 0$), and is regarded being equivalent to $\exp(\gamma_T/k_B)$ (*ca* 2) at around room temperature [93, 99]. While equation (4.57) suggests that the activation energy of the conductivity, ΔE_σ^* corresponds to the position of the Fermi level on extrapolation to $T = 0$, the actual situation is not quite so simple as this.

(b) Thermoelectric power

The thermoelectric power effect is used for identifying the conduction type in amorphous semiconductors and for studying details of the transport processes complementary to the conductivity. Corresponding to equation (4.47), the thermoelectric power $S(T)$ is given as follows [96]:

$$S(T) = (k_B/q)\frac{1}{\sigma(T)} \int \frac{E - E_F}{k_B T} \sigma_\infty(E) \left[-\frac{\partial f(E)}{\partial E}\right] dE \tag{4.58}$$

where q is the charge of the carrier. In the case of semiconductors, this relationship is written as below by using the Boltzmann approximation:

$$S(T) = (k_B/q)\left[\frac{|E_c - E_F|}{k_B T} + A\right] \tag{4.59}$$

where the factor A is given by $[1 + (\mathrm{d}\ln\sigma(E))/(\mathrm{d}\ln E)]$, and in the case of amorphous semiconductors, $A = 2$ from equation (4.51), or $A = 1$ if it is assumed that the differential conductivity $\sigma(E)$ is constant in the vicinity of the transport energy. In any case, A may be regarded as having a value of 1–2. As for the term $E_c - E_F$, some authors claim that the γ_G correction exists as in the case of the conductivity [107], while others claim that it does not exist at all [108]. The latter view will be followed for the time being.

Here, let us define a new function, $Q(T)$, in combination with equation (4.48), which takes the following form [98, 109]:

$$Q(T) = \ln[\sigma(T)] + (q/k_B)S(T)$$
$$\cong \ln[\sigma_o] + \frac{\gamma_G}{k_B} + A \qquad (4.60)$$

Thus, with the Fermi level shift and others removed, the pre-factor for the conductivity can be obtained from this relationship. A typical example of $Q(T)$ in n-type a-Si:H is illustrated in Fig. 4.20 [109]. While according to equation (4.60), $Q(T)$ shows no temperature dependence, the data presented in Fig. 4.20 indicate that there is such a dependence, as $Q(T) = Q_0 - C/k_B T$, with C increasing from *ca* 50 to 150 meV as the doping level is increased. In order to explain this phenomenon, a long-range potential fluctuation effect has been proposed [109].

As will be discussed later, it is assumed that doping with impurities creates ionized impurities and defects of inverse charge dangling bonds in approximately

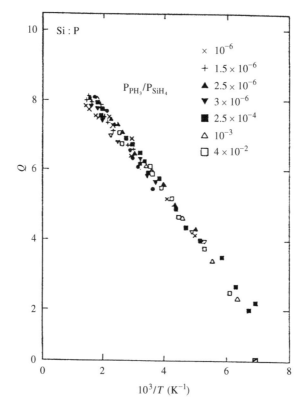

Fig. 4.20 Temperature dependence of the Q-function (see text for definition) in n-type a-Si:H [109, reproduced with permission]

equal proportions at random positions [110]. These charges may cause fluctuations in the long-range electrostatic potential, i.e. spatial fluctuations in the
transport energy E_c. The magnitude of the fluctuation (standard deviation Δ in a
Gaussian distribution) is proportional to the square root of the concentration of
the ionized impurities [98]. Numerical analysis suggests that while the potential
fluctuation does not significantly affect the conductivity [111], it does cause a
change of as much as Δ/e in the thermoelectric power [109]. The difference
is based on the presence of an $(E_c - E_F)$ factor (transported thermal energy) in
equation (4.59), which reflects a large contribution of the transport under peak
conditions of potential for thermoelectric power. By considering these factors,
equation (4.60) can be rewritten as follows [109]:

$$Q(T) \cong \ln(\sigma_0) + \frac{\gamma_G}{k_B} + A + 1.8 - 1.25\frac{\Delta}{k_B T} \qquad (4.61)$$

This relationship successfully explains the trend shown in Fig. 4.20, and if
this interpretation is correct, a potential fluctuation Δ of ca 40–120 meV should
occur in n-type a-Si:H. For p-type a-Si:H, a slightly larger value has been
reported [98].

(c) Conductivity pre-factor and mobility

In Fig. 4.20, if the value of Q obtained by the extrapolation to $1/T = 0$, i.e.
$Q_0 = 10.5$, is put into equation (4.61), together with $\gamma_G = 2.5K_B$ (for non-doped
a-Si:H), the pre-factor for electron conduction (in the conduction band) is found
to be $\sigma_o = 50$–150 S cm^{-1} [109]. In addition, if it is assumed that the Fermi
level shift is not very large, then the pre-factor derived from the temperature
dependency of the conductivity is ca 150 S cm^{-1} [112–114], and furthermore, if
the γ_T correction (which is not that well established) is added, it can be concluded
that $\sigma_o = 75$ S cm^{-1} [93]. Currently it does seem to be fairly close to the truth,
to consider the pre-factor for electron conduction to be ca 75 S cm^{-1}. For hole
conduction (in the valence band), it is generally believed that the pre-factor is not
very different from the value found for electron conduction, i.e. with in a factor
of 2–3 [109, 114], although conclusive experimental data are not available.

With the pre-factor σ_o determined, the mobility for band conduction can
be estimated from the relationship, $\mu_c = \sigma_o/ek_B TN(E_c)$. By using $N(E_c) =$
2–5×10^{21} cm^{-3} eV^{-1} [115] at room temperature, the electron mobility is calculated as $\mu_c = 10$–20 cm^2 V^{-1} s^{-1} or so. Although this value agrees with the
theoretical estimation mentioned previously in Section 4.3.1(a), this agreement
is to be regarded as being somewhat accidental, because the temperature was set
to zero in the theoretical study, as long as inelastic scattering has real importance.
From equation (4.52), the inelastic scattering length L_i in question is estimated
to be ca 10 Å at around room temperature. The validity of this value will not
be discussed further here, as a detailed description is presented in Mott's publication [93]. It is known that L_i is dependent on temperature, as T^{-s}, where s is

ca 0.6 for phonon scattering and *ca* 1 for electron–electron scattering [98, 99]. It can be pointed out, therefore, that L_i has a value of *ca* 30 Å at [98, 99] liquid nitrogen temperatures. A discussion of these physical quantities, including hole conduction in the valence band, will be given later in Section 4.4. together with interpretations of the experimental data for transient current measurement.

4.3.3 HALL EFFECT

When freely moving carriers are scattered by the incompleteness of the crystal structure, impurities or phonons, as in the case of crystal line semiconductors, the Hall coefficient, R_H, is given by

$$R_H = \frac{C}{qn} \tag{4.62}$$

As is well known, the charge sign q/e and the concentration n can be determined from this relationship. The coefficient C is a correction term which is close to 1, and depends upon the scattering mechanism. It is also well established that in a-Si:H, if the (conduction) carrier is an electron, then the Hall coefficient is positive with respect to thermoelectric power and doping, and if the carries is a hole, the coefficient is negative (the pn double anomaly) [115]. A number of arguments [117, 118] have been presented concerning this issue, including a similar anomaly in chalcogenide amorphous semiconductors (the p anomaly), but no satisfactory interpretation has been obtained as yet [119].

In connection with the previous discussions, other literature can be referred to. Here, let us examine the extension [120] of a recent idea of Mott [121], which considers an electron system described by the model Hamiltonian shown in equation (4.10). For the sake of simplicity, it is assumed that the random site energy v_i can take only one of two values; i.e. 0 or Δv. In this case, electrons in the reference crystal system are scattered only at Δv sites distributed at spacings close to the mean free path l, in order to determine the mobility edge associated with it. Let us assume that three of these Δv sites are arranged as shown in Fig. 4.21 with respect to the magnetic field H and the electrical field E. Since a classical electron (with negative charge), incident in the direction e_x, is deflected to the direction of $E \times H$, if the probability $P_3(H)$ of the electron incident at site 1 moves to site 3 increases in the presence of the magnetic field, the normal Hall effect (negative) will be observed. The three sites have been arranged in order to set the minimum conditions for identifying the direction of the electron movement.

When an electron moves from an i-site to a j-site in the presence of a magnetic field, its phase is shifted by an amount which is dependent on $k_E l$ and the magnetic field factor $eH/2c\hbar r_i \times r_j$ [117]. By considering the multiple scattering between the three sites, the change in $P_3(H)$ caused by the magnetic field is calculated as follows:

$$\Delta P_3(H) \propto -\Delta v \sin(k_E l - \phi) \tag{4.63}$$

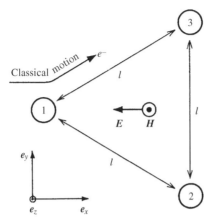

Fig. 4.21 A conceptual diagram for discussing the anomaly in the Hall coefficient in amorphous semiconductors; numbered sites separated by the mean free path l are to be regarded as scattering bodies for electrons in the crystal

where ϕ is a phase change caused by the multiple scattering, and is represented by the following:

$$\phi = \text{sgn}(\Delta v) \arctan \left(\frac{2\sqrt{\pi}}{3} k_E l \right) \tag{4.64}$$

It will be readily understood that with the mobility-edge condition (see equation (4.49)) applied, if $\Delta v < 0$, then $\Delta P_3(H) > 0$ (normal Hall effect) and if $\Delta v > 0$, then $\Delta P_3(H) < 0$ (anomalous Hall effect). In this way, if the mobility edge is determined by the scattering at $\Delta v > 0$, then the anomalous Hall effect of the electron can be readily explained. In a similar way, in the case of hole conduction, if scattering is assumed to occur at $\Delta v < 0$, then an anomalous Hall effect will be produced.

If these arguments are valid, it can be proposed that a-Si:H contains strong Si−Si bonds at concentrations of the order of 10^{20} cm^{-3}, assuming the mean free path mentioned in Section 4.2.2(b). The presence of condensed bonds and micro-void structures involving hydrogen [14] may contribute to a successful explanation. In the case of chalcogenide amorphous semiconductors, where the conduction band is regarded as being composed of anti-bonding orbitals, and the valence band composed of lone-pair (non-bonding) orbitals of chalcogens, there is no reason for the stress due to compression and expansion of localized structures exerting any reverse effects on the orbital (site) energies. It is therefore no wonder that anomalous Hall effects occur [117] only in positive-hole conduction processes.

If the concept of the mobility edge is ignored in equation (4.63), with the wavenumber k_E being replaced by the energy $k_B T$, then by using an effective mass of 0.3 m_0 and a mean atomic interval of 2.35 Å, it can be expected that the normal Hall effect will appear if the mean free path is 40–50 Å, or longer. This agrees with the experimental findings [119] that as the grain size is increased in n-type microcrystalline Si:H, the Hall sign changes from positive to negative at around 30–50 Å.

4.3.4 IMPURITY DOPING EFFECT

(a) Doping characteristics and related models

It is true to say that a-Si:H has attracted attention as an electronic material and has been widely used in various applications on account of the fact that its conductivity and conduction type (p or n) can be readily controlled by doping with impurities in the same way as crystal line semiconductors, despite its amorphous nature. Fig. 4.22 shows some experimental data of historical importance, which were originally reported by spear and coworkers at the university of Dundee [123]. It can be seen that the conductivity has been increased by several

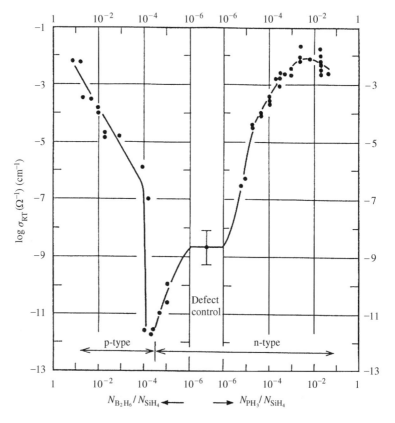

Fig. 4.22 Illustration of the control of conduction type and conductivity by impurity doping in a-Si:H [123, reproduced with permission]

orders of magnitude, for both p- and n-type doping. Since the conductivity is of the activation type, this may be attributed to a shift of the Fermi level toward either the valence-band or conduction-band edge (by 0.2 eV or so, from the band edge). The fact that the location of the conductivity minimum is slightly shifted into a p-type doping site means that the Fermi level in non-doped a-Si:H is positioned a little towards the conduction band (from the midpoint of the band gap) i.e. it has a slight n-type character. Detailed descriptions of thermoelectric power studies, changes in the optical properties and other relevant experimental data can be found in various publications by the Dundee group and other workers [124, 125]. Let us now look into the doping effects.

The doping efficiency, η, is defined as the ratio of the proportion of impurity atoms with a coordination number of four, C_4, incorporated into the film, e.g. [P$_4$]/[P], to the proportion of dopant gas, C_g, in the gas phase, e.g. [PH$_3$]/([SiH$_4$] + [PH$_3$]), as follows:

$$\eta = C_4/C_g = (C_4/C_s) \times (C_s/C_g) \tag{4.65}$$

where C_s represents the impurity concentration in the film, e.g. [P]/([Si] + [P]). The ratio C_4/C_s corresponds, therefore, to the doping efficiency η_s in the solid phase. Various doping efficiency data, as determined by a combination of the techniques used for assessing the defect density by ESR, photoluminescence and absorption spectroscopies in the sub-gap energy region, with that used for measuring the density of electrons (or holes) which fill the shallower localization levels (the so-called 'sweep-out' method [126], a form of transient current spectroscopy) are shown in Figs. 4.23 (for η_s) and 4.24 (for η) [127–129]. It is evident that each of these doping efficiency values is lower than 10%, being reduced in inverse proportion to the square root of the impurity content, with the exception of As. If the data obtained for As are excluded, it may be claimed that this relationship is universally applicable over all values of the doping efficiency. This suggests that surface reactions in the film-growth process, i.e. gas/solid interface reactions, constitute an important factor in the mechanisms of impurity doping.

While various doping models have been proposed which take into account possible chemical reaction processes [130, 131], only simple thermodynamic aspects of the solid phase will be presented here [132–134]. The following discussion will not consider the effects of hydrogen in the film [135–137], which are presumed to be of some importance, nor the microscopic reactions between impurities and defects. It is assumed that in n-type doping based on P impurities, the Fermi level does not reach the P-donor level, i.e. most of the four-coordination donor atoms are in the ionized state, P_4^+. The energy required for generating a P_4^+ state from a P_3^0 state is higher than that required for generation from the neutral donor, by an amount $(E_F - E_P)$ (where E_P is the donor level position) as the charged state has to be changed (i.e. $0 \longrightarrow +$). If thermal equilibrium is assumed, then the concentration of P_4^+ is given by the following approximation:

$$[P_4^+] \approx K \exp[(-E_F + E_{Fi})/k_B T_g][P] \tag{4.66}$$

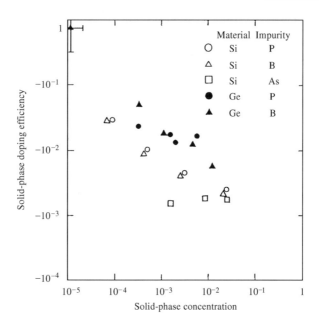

Fig. 4.23 Dependence of the solid-phase doping efficiency on the impurity content of various (film) materials [128, reproduced with permission]

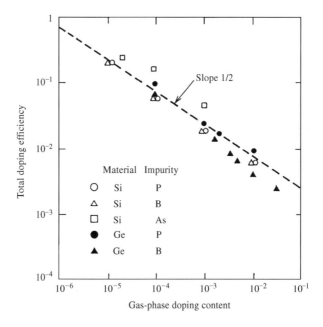

Fig. 4.24 Dependence of the total doping efficiency on the impurity content in the gas phase [128, reproduced with permission]

where K is a term which includes the energy of generation of the neutral donor, E_{Fi} is the Fermi level of non-doped a-Si:H, and T_g corresponds to growth temperature (or quench temperature) Electrons supplied from the ionized P_4^+ state raise the Fermi level. If there is thermal equilibrium between the DB defects and the underlying Si bonding states, such as weak Si–Si bonds, the following relationships are valid for the same reasons as those presented above [134]:

$$[D^-] \approx \exp[(E_F - E_{Fi})/k_B T_g][D^-]_i \qquad (4.67)$$

$$[D^0] \approx [D^0]_i \qquad (4.68)$$

$$[D^+] \approx \exp[(-E_F + E_{Fi})/k_B T_g][D^+]_i \qquad (4.69)$$

where 'i' represents the non-doped semiconductor.

Therefore, when the Fermi level is raised by P-doping, the generation of negatively charged dangling bonds is promoted, as shown in equation (4.67). If the charge neutrality condition is determined by P_4^+ and D^-, then elimination of the Fermi level from equations (4.66) and (4.67) gives the following relationship:

$$[P_4^+] \propto [D^-] \propto [P]^{1/2} \qquad (4.70)$$

This relationship indicates, therefore, that the doping efficiency ($[P_4^+]/[P]$) is inversely proportional to the square root of the impurity concentration ($[P]$).

Equation (4.70) also shows that the DB density (D^- in this case) increases in proportion to $[P]^{1/2}$, which is supported by the experimental data presented in Fig. 4.25 [127]. In this figure, the density of the electron-filled levels refers to a total summation which covers the density of the free electrons in the conduction band, the density of the electrons filling shallow levels (such as the tail level), and neutral donors, which account for $\leqslant 10\%$ of the DB density. It should be noted that the density of the free electrons comprises only a very small part of the total density. The failure to increase conductivity beyond 10^{-2} S cm^{-1} by doping in a-Si:H may be attributed to this fact, in addition to the lower mobility (in comparison to crystalline Si).

(b) Impurity states

In samples of a-Si:H doped with ≥ 100 ppm of P or As, ESR signals have been observed at low temperatures, showing hyperfine structures which are seemingly attributable to these impurities [139]. On the basis of analysis which identifies the source of these signals as being P_4^0 or As_4^0, i.e. neutral states (electron-filled) of four-coordination donor levels distributed in the vicinity of the conduction band edge, it has been claimed that, in case of P, for instance, the ratio $[P_4^0]/[P_4^+]$ is ca 10%, with most of the electrically active P atoms being ionized [129, 139]. The doping model described above is based on these experimental data, although the latter were published at a later date. The temperature dependence of the ESR signals, evaluated for either P or As, suggests that the donor level is located at about 100–150 mV below the conduction-band edge [139].

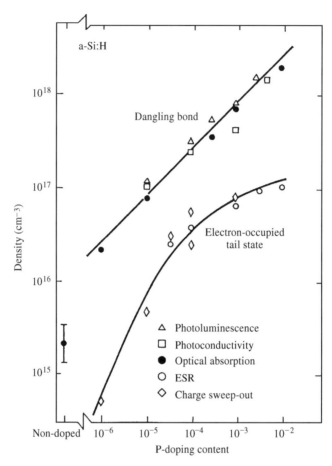

Fig. 4.25 Changes in the densities of the dangling-bond and electron-filled-tail states as a result of P-doping [127, reproduced with permission]

Some authors however, have identified the source of the hyperfine-structure ESR signals as P_2^0 [140], or electrons in the anti-bonding orbitals of weak Si−P bonds [141]. In addition, there are experimental data indicating that its energy position is located deep in the forbidden band [142]. No decisions have yet been reached as to which of these arguments is valid. Although not discussed here, changes in the solid-phase doping efficiency or conductivity have been reported which are dependent on heat treatment and the application of an electrical field [143–145], and interpretations based on the hydrogen passivation effects of donors and acceptors have been attempted [146–148]. These concepts are expected to bring forth new understandings of the doping effects, thermal equilibrium processes and photo-induced phenomena, seemingly involving interactions among hydrogen-related impurities, Si−Si bonds and defects.

4.4 Transport and Recombination of Excess Carriers

4.4.1 CARRIER CAPTURE AND NON-RADIATIVE RECOMBINATION

(a) Multi-phonon processes and capture cross-sections

The proportion of free carriers captured at a localized state (capture center) in unit time and within a unit volume is expressed by the product of the center-specific capture cross-section σ, the mean velocity of the carriers $\langle v \rangle$ and the carrier density n. Let us define $C(= \sigma \langle v \rangle)$ as the capture coefficient. Usually, the thermal velocity is used for $\langle v \rangle$, but in the case of amorphous semiconductors, $\langle v \rangle$ is defined as $(6k_B T/el)\mu^f$, where μ^f is the mobility at the band edge and l is the mean free path [149]. The mean velocity of the carriers is of the order of 10^7 cm/s^{-1} at around room temperature. This expression for the capture coefficient C is valid when a carrier approaches the capture center in a 'ballistic' fashion, i.e. the mean free path l is significantly large in comparison to $(\sigma/\pi)^{1/2}$, which is known as ballistic capture. In the case of amorphous semiconductors where the mean free path can be as small as 10 Å or so, the relationship is somewhat reversed. In this case, carriers approach the capture center in a diffusive manner, and the capture coefficient is then given by $4D^f(\pi\sigma)^{1/2}$ (diffusive capture) [150], where D^f is the diffusion length of the free carrier ($= \mu^f k_B T/e$). While it is not clear which of these expressions is more appropriate for the carrier-capture process in a-Si:H, the following discussion is based on the assumption that the process here is of a ballistic nature.

When a band-edge carrier is captured at a localized state (let us assume its depth is E_T), an energy E_T is required to dissipate it. The capture cross-section mentioned above is defined by the capture probability. If the energy is transferred into light, the capture or recombination is a radiative process (such as photoluminescence), while if it is consumed in excitation of the lattice system or other electrons, it is a non-radiative process (phonon process, or Auger process, respectively). The description given below will provide an outline of the non-radiative capture (recombination) process involving phonons [152]. If E_T is larger than the phonon energy E_{ph}, then multiple phonons are involved in the transition from the conduction-band edge to the localized state. The configuration-coordinates model shown in Fig. 4.6(a) will be useful for understanding a multiple phonon process such as this. The Huang–Rhys parameter, which is defined as $S = E_r/2E_{ph}$, where E_{ph} is the energy of the phonon concerned, serves as an index for the magnitude of the electron–phonon interaction [153]. For the sake of convenience, situations in which $S \gg 1$ and $S \ll 1$ are identified as having strong coupling and weak coupling, respectively.

Under conditions of high temperatures and strong coupling, transitions take place, by surmounting the energy barrier E_B, at the cross-over point Q_x in Fig. 4.6(a). The capture cross-section is, therefore given by the following approximation [154]:

$$\sigma(T) \approx \sigma_s \exp(-E_B/k_B T) \tag{4.71}$$

where σ_s has a maximum value of 10^{-14} cm^2, even when capture at the charge center is taken into consideration [152]. At lower temperatures (including room temperature) and with weak coupling, a transition occurs from the conduction-band edge Q_C to the localized state Q_L via a phonon-oscillation level, with $p(= E_T/E_{ph})$ phonons released. In this case, the capture cross-section has been shown to be as given in the following:

$$\sigma(T) \approx \sigma_w S^{p-1}(n_{ph} + 1)^p e^{-2n_{ph}S} \qquad (4.72)$$

where n_{ph} is the Bose–Einstein distribution function of the phonon. The temperature-independent term is written in the form of $\exp(-\gamma E_T/E_{ph})$, where $\gamma = \ln(1/S)$. Therefore, as the localized state becomes deeper, the capture cross-section becomes smaller exponentially (the energy–gap law [156]). Such a dependence on energy has been observed in carrier-capture experiments, using isothermal capacitance thermal spectroscopy (ICTS), ($D^0 + e \longrightarrow D^-$ process) with n-type a-Si:H, and the capture cross-section has been reported to be within the range 10^{-19}–10^{-17} cm^2 [52]. These data provide direct evidence for the claim mentioned in Section 4.1, i.e. that D^0/D^- is located 0.5–0.6 eV below the conduction-band edge in n-type a-Si:H (attempt-to-escape frequency; $\nu_n = \sigma\langle v \rangle N_c \sim 10^{27}\sigma$ s^{-1}). It should be mentioned that the experimental data for carrier capture since has been re-examined by considering the potential fluctuations referred to in Section 4.3 [157].

For the capture cross-sections of the dangling bonds (at around room temperature), estimated by the time-of-flight (TOF) method (to be described later), the following values have been obtained [158, 149]: $\sigma_n^+(D^+ + e \longrightarrow D^0) = (1.0 \times 10^{-15})$–$(2.7 \times 10^{-15})$ cm^2, $\sigma_n^0(D^0 + e \longrightarrow D^-) = (5.5 \times 10^{-15})$–$(1.3 \times 10^{-14})$cm^2, $\sigma_p^0(D^0 + h \longrightarrow D^+) = (1.1 \times 10^{-16})$–$(8.0 \times 10^{-15})$ cm^2, and $\sigma_p^-(D^- + h \longrightarrow D^0) = (2.7 \times 10^{-15})$–$(2.0 \times 10^{-14})$ cm^2, where it should be noted that the values given for the doped and non-doped samples are mixed. The capture cross-section of an electron, σ_n^+, has been observed to depend on $T^{-s}(s = 1$–$3)$, and by analogy with the phonon-cascade process in crystalline semiconductors [159], a capture process via a shallow localized state (e.g., a tail state) has been proposed [149]. For σ_n^0, which has been studied most frequently, a variety of theories have been proposed, including weak-coupled temperature dependence [52], dependence on $T^{-s}(s = 1$–$2)$ [119], strong-coupled temperature dependence ($E_B \sim 0.04$ eV) [159], and strong inverse activation dependence [56]. It can therefore be claimed that the complete details of the capture cross-section and capture process have not yet been established.

(b) Geminate process

Let us assume that pairs of electrons and positive holes are formed in a-Si:H through photoexcitation with an energy $\hbar\omega$ which exceeds the band gap. These carriers release phonons, which become relaxed at the respective band edges within ps intervals [161]. In this case, the relative distance between the electron and the hole (thermalization distance) is given by the following expression:

$$r_{th} \approx \left[D_r \frac{\hbar\omega - E_g + e^2/4\pi\varepsilon E_0 r_{th}}{E_{ph}\nu_{ph}} \right]^{1/2} \qquad (4.73)$$

where D_r is the diffusion coefficient for relative motion of the pairs, ε the relative dielectric constant, and ν_{ph} the phonon frequency. As can be readily seen from Fig. 4.26, if the following relationship holds:

$$e^2/4\pi\varepsilon E_0 r_{th} > k_B T \qquad (4.74)$$

an electron and a hole when produced simultaneously fail to escape thermally from the Coulomb attraction and thus recombine together immediately. This is termed geminate recombination, and must be taken into consideration for amorphous semiconductors in which the mobility or diffusion coefficient is small enough, as indicated by equation (4.73). If the distance defined by an equality for equation (4.74) is denoted by r_c, the probability of producing free carriers which can escape from geminate recombination is given by the following:

$$\eta \approx \exp(-r_c/r_{th}) \qquad (4.75)$$

This relationship applies to cases where no electrical field is present. In the presence of an electrical field, the electrons and holes are subjected to forces in

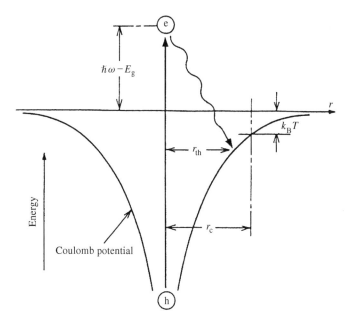

Fig. 4.26 A schematic diagram for illustrating the geminate recombination process based on Onsager's model; no electrical field is applied

mutually opposite directions, and the larger the electrical field, then the greater becomes the probability of generating free carriers [163].

It has been reported that in a-Si:H, the probability of producing free carriers is $\geqslant 90\%$, if an electric field of $\sim 10^4$ V/cm^{-1} is applied and excited with energy higher than the band gap energy, at least at around room temperature [164, 165]. This corresponds to $r_{th} = 100–400$ Å, and when $\varepsilon = 12$ (or $r_c = 46$ Å), $E_{ph} = 55$ meV, and $\nu_{ph} = 10^{12}$ s^{-1} are put into equation (4.73), the mobility, $D_r e/k_B T$, is estimated as ~ 100 cm^2/V^{-1}s^{-1}. This value is greater by an order of magnitude than the estimated free carrier mobility at the band edge (10 cm^2/V^{-1}s^{-1}) (to be discussed later). The higher mobility may be attributed to the higher energy at the band edge, but measurements of fast relaxation processes for photo excited carriers rules out this possibility [166]. The model for the geminate process described here is a somewhat classical one, and in an the actual system, the relative motions of the electron and holes, for example,are quantized (involving excitons). It is therefore of particular interest to see what might result if the concept of excitons is taken into consideration.

4.4.2 RECOMBINATION AND PHOTOCONDUCTIVITY UNDER STEADY ILLUMINATION CONDITIONS

(a) Non-thermal equilibrium statistics for localized states

In order to describe the behavior of carriers under non-thermal equilibrium conditions such as optical excitation, it is essential to know the electron occupation probability for the localized state which serves as a trap or a recombination center. Let us first study the non-thermal equilibrium steady distribution function for a correlated defect (three-state, two-level system) such as the dangling bond (DB). With reference to Fig. 4.27, the probability (distribution function) F^μ for the DB being in the state D^μ ($\mu = +, 0$, or $-$) is readily obtained from the following [167]:

$$F^+ = P^- P^0 / (N^0 N^+ + N^+ P^- + P^- P^0) \qquad (4.76)$$

$$F^0 = N^+ P^- / (N^0 N^+ + N^+ P^- + P^- P^0) \qquad (4.77)$$

$$F^- = N^0 N^+ / (N^0 N^+ + N^+ P^- + P^- P^0) \qquad (4.78)$$

Assuming that the DBs form two levels each, i.e. D^+/D^0 and D^0/D^-, at energies $E(+/0)(= E)$ and $E(0/-)(= E + U_{\text{eff}})$, respectively the parameters in equations (4.76)–(4.78) are given as follows:

$$N^0 = C_n^0 n + e_p^0; e_p^0 = (C_p^- N_v) \exp[(E_v - E - U_{\text{eff}})/k_B T] \qquad (4.79)$$

$$N^+ = C_n^+ n + e_p^+; e_p^+ = (2C_p^0 N_v) \exp[(E_v - E)/k_B T] \qquad (4.80)$$

$$P^0 = C_p^0 p + e_n^0; e_n^0 = (C_n^+ N_c/2) \exp[(E_v - E_c)/k_B T] \qquad (4.81)$$

$$P^- = C_p^- p + e_n^-; e_n^- = (2C_n^0 N_c) \exp[(E + U_{\text{eff}} - E_c)/k_B T] \qquad (4.82)$$

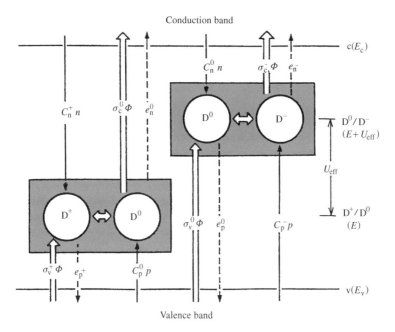

Fig. 4.27 Illustration of various carrier transition processes involving dangling bonds where arrowed solid lines represent capture processes, broken lines thermal emission processes, and bold lines optical transition processes

where C_i^μ is the carrier capture coefficient described in Section 4.3, and e_i^μ corresponds to the carrier thermal emission rate ($\mu = +, 0,$ or $-$, and $i = n$ or p). The terms, E_c, and N_c, and E_v and N_v, denote the energy position and effective state density, at the conduction-band edge and valence-band edge, respectively. It is evident that if n and p, as well as their counterparts at thermal equilibrium, i.e. n_0 and p_0, are used, then the DB distribution functions under thermal equilibrium conditions can be obtained by a simpler calculation, as shown previously by equations (4.17)–(4.20).

If such DBs exist at a density N_D, the carrier recombination factor R (via DBs in the steady state) is given by the following [167]:

$$R = \frac{(np - n_i^2)(C_n^0 C_p^- N^+ + C_p^0 C_n^+ P^-)}{N^0 N^+ + N^+ P^- + P^- P^0} N_D \qquad (4.83)$$

where n_i is naturally equal to $(N_c N_v)^{1/2} \exp(-(E_c - E_v)/2k_B T)$. The charge density due to the DBs, is represented by $Q = e(F^+ - F^-)N_D$. If the level energy is distributed, then expressions for R and Q have to include a summation or an integral over these states.

For the case of a non-correlated two-state, one-level system, where the charged state can only take a value of 0 and ± 1, Fermi statistics or Shockley–Read-type statistics [168], under non-thermal-equilibrium, steady-state conditions, can be used as they stand. This situation can be handled by removing an irrelevant term in our description of the correlated system, involving the D^+/D^0 or D^0/D^- levels; details are not presented here because they can be found in any standard text book on semiconductor physics. However, let us at this point give an outline of Simmons–Taylor model [169] which extends these statistics to a localized-state system having an energy distribution. The electron occupation factor (distribution function), $f(E; n, p)$, under non-thermal-equilibrium, steady-state conditions for a set of localized states, where the capture coefficients for the electrons and holes, C_n and C_p, respectively, are constant, i.e. independent of energy, is as shown in Fig. 4.28 ($np > n_i^2$). Three characteristic energy values are each defined for the electrons and holes. In the case of the electron, for example, these are the quasi-Fermi level of the band electron, (E_{Fn}), the trap Fermi level for determining the electron-occupation factor for the localized state, (E_{tn}), and the demarcation level (E_{dn}), which is defined by $e_n = C_p p$. These are represented by equations (4.84)–(4.86), respectively, as follows:

$$E_{Fn} = E_c - k_B T \ln[N_c/n] \tag{4.84}$$

$$E_{tn} = E_c - k_B T \ln\left(\frac{C_n N_c}{C_n n + C_p p}\right) \tag{4.85}$$

$$E_{dn} = E_c - k_B T \ln[C_n N_c/C_p p] \tag{4.86}$$

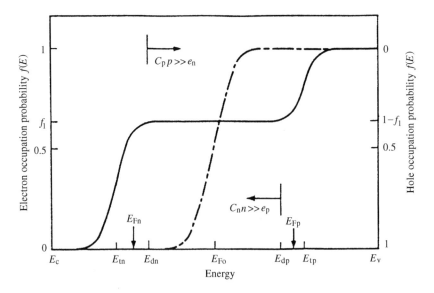

Fig. 4.28 Occupation factors of electrons (f) and positive holes ($1-f$) for the localized levels in the Simmons–Taylor model under non-thermal-equilibrium, steady-state conditions; broken line shows the Fermi distribution under thermal equilibrium conditions

For the Fermi level, the occupation factor for localized states which exist between the gap mid-point and the conduction band (i.e. $C_n n \gg e_p$, $E > E_{dp}$) can be approximated by the following expressions:

$$f(E) \approx f_1 \frac{1}{1 + \exp[(E - E_{tn})/k_B T]}; \quad f_1 = \frac{C_n n}{C_n n + C_p p} \tag{4.87}$$

If $C_n n$, $e_n \gg C_p p$, then the occupation factor of the localized states which meet this condition is determined only by the quasi-Fermi level of the electron, and the localized state is in thermal equilibrium with the band. The localized state serves as a trap for electrons, and not as a recombination center. A similar argument applies to the localized states on the side of the valence-electron band ($C_p p \gg e_n$; $E < E_{dn}$).

At lower temperatures, the recombination factor, by considering the localized states can be approximated by the following expressions:

$$R \approx \frac{C_n C_p (np - n_i^2)}{C_n n + C_p p} \int_{E_{tp}}^{E_{tn}} N(E) \, dE \tag{4.88}$$

It goes without saying that if there exist localized states with different sets of C_n and C_p parameters (acceptor-like C_n^0 and C_p^-, and donor-like C_n^+ and C_p^0, for example), then summation is required for all of these sets. Attempts have been made to apply the concept of the Simmons–Taylor approach to the correlated localized state system mentioned above [170], but it is not widely used at the present time.

(b) Continuous-Wave-Light photoconductivity

The photoconductivity, σ_{ph}, under steady-light-illumination conditions (volume carrier generation factor, g) is given by the following expression:

$$\sigma_{ph} = e(\mu_n^f \Delta n + \mu_p^f \Delta p) \propto g^\gamma \exp(-W/k_B T) \tag{4.89}$$

where Δ denotes the excess carrier component, and μ_f indicates the mobility of the band (i.e. free) carrier. In these experiments, the photoconductivity is generally considered to be proportional to the γth power of the light intensity, and its temperature dependence involves an activation energy W, as shown on the right-hand side of equation (4.89). Typical experimental data that have been obtained [124, 171–174] are as follows:

(i) If $\sigma_{ph} > \sigma_{dark}$, then W is positive to lower temperatures. As the Fermi level is shifted toward the band edge by doping, it will become smaller.

(ii) The power factor γ is reduced continuously from 1 to 0.5 (or even smaller) as the Fermi level is shifted towards the band edge.

(iii) The absolute value of σ_{ph} grows as the Fermi level approaches the conduction-band edge (n-type), for example, as shown in Fig. 4.29 [124]. (In this figure, the decrease in g towards the extreme n-type region is attributed to an increase in the DB density as a result of doping.)

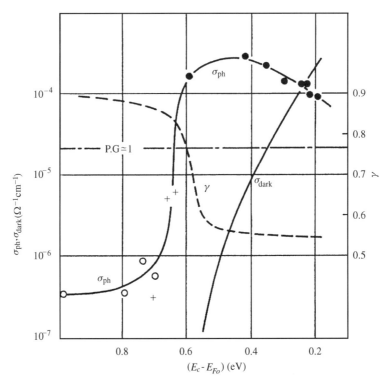

Fig. 4.29 Fermi-level dependence of the dark conductivity (σ_{dark}) and photoconductivity (σ_{ph}), and its associated intensity-dependence factor (γ) [124, reproduced with permission]

Regarding the photoconductivity, a variety of interpretations have been proposed, such as assuming correlated DBs, as referred to in Section 4.4.2(a) [167, 175, 176] the consideration of distributed non-correlated localized states [177], etc. Taking an extreme viewpoint, it may be concluded that steady-light-illumination photoconductivity can be explained in many ways, as long as a number of appropriate parameters are used. It must be admitted, however, that certain useful information on the basic physical parameters, such as the position, distribution and capture coefficient of the DB state, for example, cannot be derived by using these arguments. Therefore, let us describe here a few examples of simple interpretations of the photoconductivity. Let us first assume that the recombination occurs via a single DB which is positioned closer to the conduction band (from the middle of the bond gap). As stated in Section 4.4.2(a), the recombination factor R, through DBs, and the charge Q are expressed by using n and p only. Therefore, by taking the case where $R = g$ (steady-excitation condition) and combining this with the situation where $Q/e + p - n = 0$ (charge-neutrality condition), the photoconductivity Δn can be obtained as a function of the light intensity g and temperature T [167, 177]:

$$g \approx C^* N_D \frac{(\Delta n + n_0)\Delta n}{n_1 + n_0} \tag{4.90}$$

where C^* is the effective capture coefficient. While C^* is equal to C_n^+ when the light intensity is low, it becomes equal to $C_n^+ C_p^- / 2C_p^0$, or C_n^0, as the intensity increases; n_1 represents the electron density when the Fermi level is at D^0/D^-. It is evident from equation (4.90) that as the light intensity grows, the γ-value decreases from 1 to 0.5, and that as the Fermi level approaches the conduction-band edge, or n_0 grows beyond n_1, then Δn increases, while the g-value and activation energy for Δn both decrease. For example, the increment of Δn may be interpreted as a result of a suppression of the recombination process via D^+/D^0 level, due to the change of the DB state from D^0 to D^-.

Regrettably, this approach is inadequate for explaining g-values other than 1 or 0.5, and in particular, those smaller than 0.5. In order to cope with this, a shallow trap level must be considered. For example, let us assume that the exponential tail shown in equation (4.91) is in thermal equilibrium with the band electron (see previous section). Then, the trap electron density (equation (4.92)) can be derived as follows:

$$N_{ct}(E) = \frac{N_{ct}}{k_B T_c} \exp[(E - E_c)/k_B T_c] \tag{4.91}$$

$$n_t \cong N_{ct}[\alpha_c \pi / \sin(\alpha_c \pi)](n/N_c)^{\alpha_c}; \alpha_c = T/T_c < 1 \tag{4.92}$$

If this relationship is combined with the condition of charge neutrality, the following expression can be obtained [167, 177]:

$$1/\gamma = \frac{\partial \ln g}{\partial \ln \Delta n} \approx 1 + \frac{\Delta n}{n + n_1} + \alpha_c \frac{\Delta n}{n} \tag{4.93}$$

From this expression, it may be supposed that the γ-value changes continuously from 1, via $1/(\alpha_c + 1)$, to $1/(\alpha_c + 2)$, through the shift of the 'false' Fermi level to the conduction-band edge (high light intensity and large n_0). As the quasi-Fermi level is fully included in the tail state, the recombination will have a greater contribution, and thus the γ-value returns to $1/(\alpha_c + 1)$ [179]. It has been attempted to interpret the steady-state photoconductivity of a-Si:H on the basis of a simple model, but it should be noted that this interpretation is neither exhaustive nor exact.

4.4.3 CARRIER TRANSPORT CHARACTERISTICS UNDER TRANSIENT ILLUMINATION CONDITIONS

(a) Trap-controlled conduction

Let us consider the steady state first. As long as only the electron transport is taken into consideration, the conductivity is given by the term $e\mu^f n$. Let us assume a shallow trap satisfying the condition $e_n \gg C_p p$, and denote the electron density which fills it by with n_t. The conductivity can then be written formally as $e\mu(n + n_t)$, where μ represents the mobility of all of the carriers around the band edge, i.e. both free and trapped carriers; this is generally termed the drift mobility and is defined as follows:

$$\mu = \mu^f \frac{n}{n + n_t} \tag{4.94}$$

Since n_t is a function of n alone, as shown in equation (4.92), providing that $C_n n \gg C_p p$, then μ is also a function of n alone. This means that the drift mobility is smaller than that of the free carrier, and generally depends upon n. Sometimes, it can be very convenient to describe the carrier transport in the steady state from such a viewpoint [180–182]. On the other hand, in the case of the transient processes to be described below, this will concern the nature of the outcome.

The macroscopic transport behavior of for free electrons in a one-dimensional model $(0 < x < d)$, can be represented by the following equation:

$$\frac{\partial n(t, x)}{\partial t} \cong \int_0^t n(\xi, x)W(t - \xi)d\xi - n(t, x)/\tau^f - \mu^f E \frac{\partial n(t, x)}{\partial x} + g(t, x) \tag{4.95}$$

where the diffusion term is ignored, and it is assumed that the electron occupation probability for a shallow trap is far smaller than one, a uniform electric field E is applied, and a recombination lifetime (or deep-trapping time) τ^f, is formally used, for example, as $n/R(n, p)$. A term including $W(t)$ represents the journey of the electron between a shallow trap and a band (multiple-trapping process). In the Laplace-transform domain, $W(t)$ is defined as follows:

$$\widehat{W}(s) = -\sum_i \frac{C_i s}{s + e_i} \tag{4.96}$$

The summation is to be made for trap level i, and if the distribution is continuous, it should be replaced with an integral involving a state-density function. When the state density $N(E)$ changes more slowly than the Fermi distribution, $W(t)$ can be approximated as below by using this relationship (equation (4.96)):

$$W(t) \approx k_B T C N(E_t)/t; \quad t > 0 \tag{4.97}$$

where the capture factor C is assumed to be a constant, and the energy E_t is defined as below by using $e(E_t)t = 1$:

$$E_t = E_c - k_B T \ln(\nu t); \nu = C N_c \tag{4.98}$$

As is evident from equation (4.96), with s replaced by $1/t$, a level deeper than E_t serves as a deep trap, and only levels around it will exchange electrons with a band. (Shallower levels have already released electrons.) In this sense, E_t is called the thermalization energy or the demarcation level (in the time domain).

The current density $J(t)$ that is observed is generally given (with the diffusion current ignored) by the following:

$$J(t) = \frac{e}{d} \int_0^d \mu^f E n(t, x) dx - C_g \frac{dV(0, d)}{dt} \tag{4.99}$$

where the second term is a displacement current component, C_g the electrostatic capacitance which depends upon the sample geometry, and $V(0, d)$ is the potential difference across the sample. The following discussion is based on the assumption that the time change is negligible. In the case of the time-of-flight (TOF) measurements shown in Fig. 4.30, the current waveform can be obtained by the Laplace transform of equation (4.99) as follows:

$$\hat{J}(s) = \frac{Q_0}{t_{T^f}} \int_0^{t_T^f} \exp[-(s - \hat{W}(s) + 1/\tau^f)u] du \tag{4.100}$$

where Q_0 is the face density of the charge fed into the sample at $x = 0$ and $t = 0$, and t_{T^f} is the transit time for a free carrier $(= d/\mu^f E)$.

If E_t has shifted to a deeper place than the assumed trap-level distribution in the observed time domain, then the Laplace transform s in the denominator of equation (4.96) can be ignored, and the current waveform is given by using a step function $\Theta(t)$, as shown in the following:

$$J(t) = \frac{Q_0}{t_T} \exp(-t/\tau)[\Theta(t) - \Theta(t - t_T)] \tag{4.101}$$

where t_T is the transport time for all of the carriers (including trap carriers) and corresponds to the drift mobility $\mu = d/t_T E$, and τ is the lifetime of all of

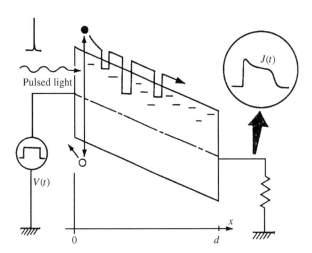

Fig. 4.30 Illustration of the measurements required in the time-of-flight (TOF) technique

the carriers. These parameters are related to the free-carrier parameters in the following way:

$$\mu = \mu^f/(1+\theta); \tau = \tau^f(1+\theta) \tag{4.102}$$

where:

$$\theta = \sum_i C_i/e_i \approx (n_t/n)_0 \tag{4.103}$$

This subscript 0 indicates the value for thermal equilibrium. Equation (4.102), in fact, agrees with the more intuitive expression given by equation (4.94). It should be noted, however, that under the conditions that apply to this latter case, the value of $\mu\tau$ obtained by the TOF measurements agrees with that obtained for the free carrier, i.e. $\mu^f\tau^f$.

(b) Anomalously dispersive conduction

In the examples described up until now, as the diffusion current component is ignored, then the distribution of the free carriers is in the form of a delta function (which should be a Gaussian function otherwise) and the position of its center of gravity $\langle x \rangle$ is proportional to the time t. However, trap levels are distributed over a broad energy range, and if E_t crosses a finite $N(E)$ within the observed a time span then, the distribution of free carriers may deviate widely from a delta or a Gaussian function, even if the diffusion-current component is negligibly small. Naturally, the current waveform will not be as simple as that shown in equation (4.101). This phenomenon is known as anomalous-dispersion-type conduction, and is based on a multiple-trapping mechanism [183–186]. Let us try here to produce an approximate representation of this situation. In equation (4.100), a new function, namely:

$$\hat{W}_0 = \hat{W} + \sum_i C_i$$

is defined to confine the effects of the deeper trap into the second term, the following approximations will be obtained:

$$\hat{J} \approx \frac{t_{T^f} Q_0}{2} W_0 \quad \text{(for the longer-time region)} \tag{4.104}$$

$$\hat{J} \approx -\frac{Q_0}{t_{T^f}} W_0^{-1} \quad \text{(for the shorter-time region)} \tag{4.105}$$

If an exponential form of equation (4.91) is assumed for the trap-level distribution, we obtain $\hat{W}_0 \propto -s^{\alpha_c}$. If this relationship is put into equations (4.104) and (4.105), current waveforms which are proportional to $t^{-1-\alpha_c}$ (for the longer-time domain) and to $t^{1-\alpha_c}$ (for the shorter-time domain) are obtained. The waveform for the longer-time domain may be represented as $J(t) \propto N(E_t)/t$ by putting equation (4.97) into equation (4.104). For the shorter-time domain, the waveform may be approximated by $J(t) \propto 1/N(E_t)t$ [187], but application of the latter should be treated with caution [188].

Before the leading edge of the charge flux has crossed a sample, all of the injected charge remains in the material. Hence, the time-dependent mobility is defined by $J(t) = \mu(t)EQ_0/d$. This is proportional to $t^{1-\alpha_c}$, and the apparent charge transport time t_T^*, which may be regarded as the transition between the above mentioned two current responses, is given by the following if $T < T_c(\alpha_c < 1)$, where the coefficients are fully specified:

$$t_T^* \cong \left[\frac{d}{\mu^f E}\right]^{1/\alpha_c} v^{1/\alpha_c - 1} \left[\left(\frac{N_{ct}}{N_c}\right)\left(\frac{\alpha_c^2 \pi \Gamma(\alpha_c)}{\sin \alpha_c \pi}\right)\right]^{1/\alpha_c} \tag{4.106}$$

If the drift mobility is calculated from this expression, the result will depend upon both the electric field and sample thickness, and hence cannot be a physical parameter which is genuinely characteristic of the material. The position of the center of gravity of the free carrier $\langle x \rangle$ is proportional to t^{α_c} for t_T^*, and its distribution is of the non-Gaussian type, with the maximum set at around the excited surface [150]. It should be added that such an anomalously dispersive conduction behavior does not correspond directly to the exponential distribution of the localized states, but inevitably occurs whenever E_t crosses a finite $N(E)$.

As illustrated in Fig. 4.31, conduction of an anomalous-dispersion type becomes distinct in the lower-temperature regions, and in a-Si:H it occurs between 80 and 150 K for electron transport [186, 189, 190], and between 200 and 360 K for hole transport [186, 189, 191]. At temperatures than these conduction of a normal dispersion type (described in Section 4.4.3(a) is recovered, and the drift mobility increases [192, 193]. This may be interpreted in terms of a change in the major conduction mechanism from one of a multiple trapping kind to one of a phonon-mediated tunnelling type [194], which may occur selectively for localized states of certain depth [195, 196].

4.4.4 MOBILITY AND LIFETIME OF CARRIERS

(a) Mobility-lifetime product

When discussing the mobility (μ) and lifetime (τ), or their product ($\mu\tau$), obtained by various physical methods, it is necessary to pay attention to the fact that these are not physical parameters which are specific to the particular material, being considered but are variables which are dependent upon the various environmental and measurement conditions, such as temperature, excitation conditions, applied electric field, observation time domain and sample-electrode configuration. Specific physical parameters refer, for example to the mobility of the free carrier μ^f, the energy distribution of the localized states, and the carrier capture cross-section, while mobility and lifetime both represent certain aspects of the physical phenomena resulting from a complicated entanglement of such specific parameters.

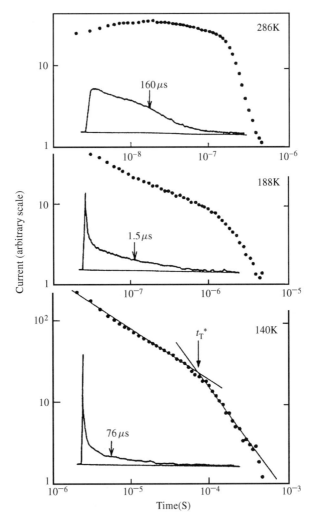

Fig. 4.31 Electron-transport response times at different temperatures obtained by the time-of-flight (TOF) technique [186, reproduced with permission]

The typical method used for $\mu\tau$ product measurement is the time-of-flight (TOF) technique where the following conditions need to be met: (i) no charge injection occurs, except for photoexcitation at one end of the sample (blocking structure), and (ii) the injected charge Q_0 is smaller than the product of C_g and the applied voltage, while the measurement time is within the dielectric relaxation time $\tau_d = \varepsilon\varepsilon_0\sigma_0^{-1}$ (field uniformity) [197]. Under ideal conditions, it is assumed that in single-carrier transport, the occupation probability of the trap level within the sample is equal to that which exists at thermal equilibrium, where changes in the occupation probability in the depth direction of the junction being used are regarded as negligible. In such a simple situation, the trapped charge Q is equal to the integral of $J(t)$, i.e. with the Laplace variable s set to zero in equation (4.100), as follows:

$$Q/Q_0 = \frac{\mu^f \tau^f E}{d} \left[1 - \exp\left(-\frac{d}{\mu^f \tau^f E} \right) \right] \tag{4.107}$$

This expression is known as Hecht's relationship [198], from which the $\mu\tau$ product of the free carrier can be derived. However, if the trapped charge is integrated up to the time domain where electron release from the deeper localized states becomes significant (even within the dielectric relaxation time), the meaning of the deep trapping time, τ^f, is lost, and equation (4.107) thus becomes invalid.

Some examples of measurements made by using this procedure are shown in Fig. 4.32 [128]. In this figure, correlations of the $\mu\tau$ products for both electrons and holes are illustrated. The proportionality, $\mu_n^f \tau_p^f \propto \mu_p^f \tau_p^f$, in non-doped a-Si:H indicates that the two carriers have a deep trap (capture center) in common. Since the $\mu\tau$ product is inversely proportional to the spin density for $g = 2.0055$, this center appears to be a neutral DB, i.e. D^0. For interpreting the correlation of the $\mu\tau$ product in weakly doped samples, equations (4.67)–(4.69) (given in

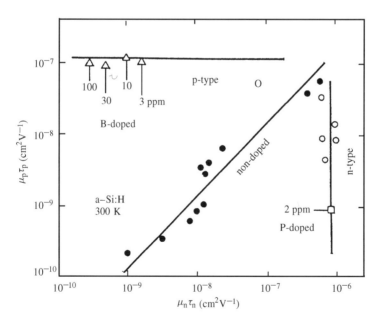

Fig. 4.32 Correlation of mobility–lifetime ($\mu\tau$) products for electrons and positive holes; open circles represent data obtained for samples containing P or B as residual impurities [23, reproduced with permission. Copyright 1983 by the American Institute of Physics]

Section 4.3) are used to derive the following expressions:

$$\tau_n^f \cong \left(C_n^0[D^0] + C_n^+[D^+]\right)^{-1} \tag{4.108}$$

$$\tau_p^f \cong \left(C_p^0[D^0] + C_p^-[D^-]\right)^{-1} \tag{4.109}$$

Therefore, in the case of an n-type sample, for example, τ_n^f fixed as $\tau_n^f \sim (C_n^0[D^f]_i)^{-1}$, and τ_p^f decreases with increased doping, as $\tau_p^f \sim (C_p^-[D^-])^{-1} \infty [P]^{-1/2}$. If a mobility value for the free carrier is postulated, it is possible to estimate the capture cross-section of the DB, as discussed in Section 4.3.1(a).

The $\mu\tau$ product can also be derived from steady-state photoconductivity measurements of σ_{ph}/eg. Values of around 10^{-5}–10^{-4} cm^2/V^{-1} have been obtained for non-doped a-Si:H of solar-cell grade. These values are two to four orders of magnitude greater than those obtained from TOF measurements, and there various arguments concerning the reasons for this [199, 200]. However, since steady-state photoconductivity results from a number of complicated processes, such as the capture, release and recombination of carriers in the presence of both electrons and holes, (as described in Section 4.4.2), it is somewhat unreasonable to compare data obtained in this way with the results from the TOF method, which is based on an entirely different measurement background. The view of this present author is that the $\mu\tau$ products derived from the steady state photoconductivity method are essentially unsuitable for the evaluation of film properties or transport processes.

The situation can be slightly improved if the photoconductivity response to a weak reference light a 'small' signal is measured while the occupation rate of the localized state is fixed by using intensively biased illumination (with a pre-reference set-up), instead of examining the steady-state photoconductivity with 'large' signals. This technique can be extended to localized-state spectroscopy, which is based on the frequency response of the photoconductivity [55, 157]. In addition, a new technique, namely steady-state photocarrier grating spectroscopy has recently been developed; this estimates the ambipolar diffusion length L_{amb}, based on changes in the photoconductivity response caused by interference between the biased and reference light sources [201]. The value of L_{amb} is ca 10^3 Å for biased illumination of 100 mW cm^2 in non-doped a-Si:H of solar-cell grade (see Fig. 4.33 [182]). This value is almost identical to the diffusion length of positive hole, L_{dp}, while the diffusion length of the electron, L_{dn}, is found to be two to three times large than this.

The parameters that can be obtained by this method are defined as shown in the following:

$$L_{amb} = \left[\frac{\mu_n L_{dp}^2 + \mu_p L_{dn}^2}{\mu_n + \mu_p}\right]^{1/2} \tag{4.110}$$

$$L_{di} = [D_i^* \tau^*]^{1/2}, \quad \text{for } i = n, p \tag{4.111}$$

$$1/\tau^* = \frac{\partial R}{\partial N} + \frac{\partial R}{\partial P}; D_n^* = \frac{\partial n}{\partial N} D_n^f; D_p^* = \frac{\partial p}{\partial P} D_p^f \tag{4.112}$$

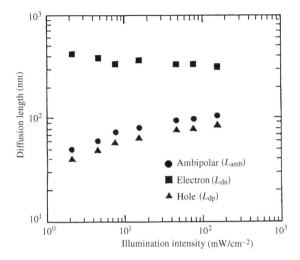

Fig. 4.33 Dependence of the ambipolar (L_{amp}), electron (L_{dn}) and hole (L_{dp}) diffusion lengths in non-doped a-Si:H on illumination intensity [182, reproduced with permission. Copyright 1992 by the American Physical Society]

where μ_n and μ_p are the ordinary drift mobility parameters defined by equation (4.94). The parameter N represents $n + n_t$, and the derivatives in these expressions give values for the charge density under biased illumination, while the parameters with an asterisk denote those obtained using small signals. As should be evident from the description so far, the diffusion lengths obtained by this method should not be compared directly with the $\mu\tau$ products derived from TOF measurements (in contrast to the photoconductivity data). However, this technique has the merit of being able to characterize solar cells, for example, under actual operating conditions, as it determines the transport parameters under biased illumination. It should be noted that the transport properties can be evaluated by using these diffusion lengths, only when the electric field is around its lower limit and electrical neutrality is maintained.

In addition, there are a number of other ways for assessing these parameters, such as measurements of the surface photovoltage (for diffusion length) [202, 203], and analysis of the dependence of solar-cell collection efficiency spectra on applied voltage (for the effective $\mu\tau$ product) [204]. As in the cases mentioned above, close attention should be paid as to how they are related to the basic physical parameters, and how useful they are for evaluating which particular parameter under specific environmental conditions [177, 182, 205].

(b) Mobility

In General, the mobility is obtained from the drift mobility parameter $\mu(= d/Et_T)$, from the transit times in TOF measurements. Electrons and holes both have conductivities of the normal dispersion type at temperatures higher than 150 and 360 K, respectively, and provide values which are independent of sample thickness and applied field [195, 206]. The drift mobility at around room temperature is of the order of 1 cm^2 V^{-1}s^{-1} (with activation energies of *ca* 0.14 eV) for electrons, and $10^{-3} - 10^{-2}$ cm^2 V^{-1}s^{-1} for holes (for reference only, owing to anomalous-dispersion behavior at this temperature). The mobilities of the free carriers can be estimated from equations (4.102) or (4.106), based on the multiple-trapping mechanism, by considering the dependence on temperature and applied field, so as to be self-consistent with the distribution of the trap levels in the vicinity of the band edge. It is a well established concept at the present time that the conduction-band edge slowly decreases to a depth of *ca* 0.15 eV, and then falls steeply after this point [195, 196, 206]. The band mobility of the electron is estimated to be *ca* 10 cm^2 V^{-1}s^{-1} at around room temperature. The variation with temperature of the valence-band edge is said to take either the form of an exponential function ($T_V \sim 500$ K) [186] or shows a decline (in the form of a Gaussian function) for a range of depth of 0.2–0.4 eV [187, 191]. Estimations of the band mobility of the positive hole have given either *ca* 0.7 cm^2 V^{-1}s^{-1} at around room temperature [186, 191], or a figure which is close to that of the electron [187]. The former value seems to be more probable.

The TOF method cannot be applied to doped samples of lower conductivity because of limitations in the measurements. As alternatives to this, either a travelling-wave method [207] or a combined conductivity 'sweep-out' technique [208] can be used. The former technique yields mobility values for n-type a-Si:H which are almost the same as those obtained from TOF measurements on non-doped samples, while the latter procedure gives values which are reduced by an order of magnitude, with increased activation energies. This may be attributed to the rise in the mobility edge caused by the potential fluctuations mentioned above [208], but it seems more plausible to consider that the spatial distribution under the more classical potential fluctuation of the free and trapped charges is responsible [209]. While these arguments lead to the conclusion that the mobility of the band electron, for example, is *ca* 10 cm^2 V^{-1}s^{-1} [210], some authors still claim (or have claimed in the past) that the band mobility is greater than 100 cm^2 V^{-1}s^{-1}. Those readers who have an interest in these disputes are recommended to read the references concerned [211].

(c) Lifetime

As mentioned at the beginning of this section, the lifetime of a camel is not a physical parameter which is specific to a particular material. It is necessary, therefore, to recognize how this parameter has been defined, and in particular, which measurement environment we are considering. Typical lifetime data, i.e. τ^f parameters, are obtained from a combination of $\mu^f\tau^f$ values derived from TOF measurements, with estimated μ^f and τ (lifetimes corresponding to drift mobilities) values directly obtained by the delayed field (DF) method [212],

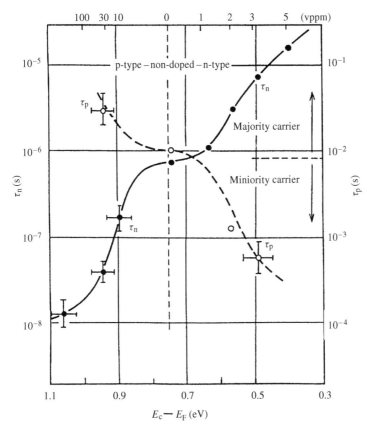

Fig. 4.34 Dependence of the lifetimes of the electron (τ_n) and hole (τ_p) on the Fermi level [213, reproduced with permission]

with the latter being a modified TOF procedure. These values may be considered as referring to the lifetimes involved in capturing excess carriers either in the band or in the vicinity of the band edge approaching the deep levels, for a single-carrier, thermal-equilibrium situation. Fig. 4.34 shows the dependence of the lifetimes of electrons and holes on the Fermi level at around room temperature, as measured by the DF method [213]. (The data shown for hole conductivity is to be regarded only as a reference because of its anomalously dispersive nature.) These trends have been interpreted in terms of changes in the charged states of the dangling bonds according to equations (4.108) and (4.17), although we will not elaborate further on this here. The lifetime of the free electrons in undoped samples, τ_n^f, has been estimated as *ca* 10^{-7} s, based on equation (4.102), while that of the free hole, τ_p^f, is estimated to be of the order of 10^{-6} s, although some authors have doubts concerning the velocity of this latter value.

It is also possible to estimate the lifetimes from photoconductivity measurements. In this case, the phase delay ϕ in the photocurrent response for modulated reference light (small signals) under illumination with an intensively biased light source is determined as a function of the modulation frequency f, with the small-signal lifetime (defined in equation (4.112) in Section 4.4.4(a)) being obtained from the following equation:

$$\tan \phi = 2\pi f \tau^* \qquad (4.113)$$

It has been reported, for example, that the lifetime τ^* in non-doped samples at around room temperature under a uniform illumination of *ca* 100 mW cm^{-2} is of the order of 10^{-6} s. [182]. The value for the lifetime is generally increased as the Fermi level is shifted by doping. While this trend is similar to the behavior of the majority-carrier lifetime shown in Fig. 4.34, it should be noted that the two should be directly compared, since both the excitation environment for the measurement, and the definition of the lifetime itself are different.

4.4.5 PHOTOLUMINESCENCE

(a) Transitions between band edges

Electron–hole pairs produced by photoexcitation are annihilated in most cases through non-radiative recombination via localized states in the higher temperature range (including room temperature), but at lower temperatures, radiative recombination involving light emission becomes more prominent. In non-doped a-Si:H, broad photoluminescence (emission) spectra with FWHM of 0.2–0.3 eV and peaks at 1.2–1.4 eV are observed, as illustrated in Fig. 4.35 [215]. Since this emission band is located at the feet of the adsorption spectra shown in Fig. 4.9, and the intensity increases in the lower temperature range (becoming saturated at around 50 K), this phenomenon is attributed to the optical transitions between localized states at the band edge (including the band tail).

In the lower-temperature range (e.g. <50 K), the electron–hole pairs generated by photoexcitation are subjected to relaxation at the band edge and shallower localized states over a very short space of time, to be fixed spatially (or localized). It is generally believed that if the excitation intensity is not that high, i.e. the mean spacing between the electron–hole pairs is not smaller than r_{th} in Fig. 4.26, then geminate recombination occurs, thus resulting in photoluminescence in the radiative process. The emission intensity (or emission efficiency) is defined on the basis of the competition with non-radiative recombination processes, e.g. via DBs. Since the probability of electron–hole pairs being thermally excited in each band is very small at lower temperatures, it is claimed that the DB-mediated recombination results from a tunnelling process from the localized states involved in the radiative process to the DBs. According to this concept, it may be shown that the emission efficiency η_{PL} at lower temperatures decays exponentially as the DB density increases [216]. In fact, such a trend has been confirmed experimentally. At higher temperatures, photo-induced electron–hole pairs are

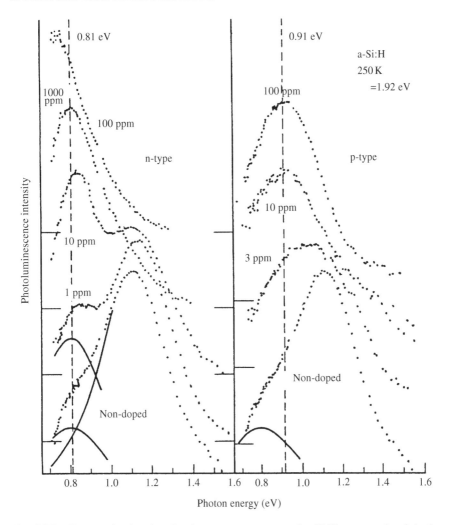

Fig. 4.35 Changes in the photoluminescence spectrum of a-Si:H as a result of doping [215, reproduced with permission. Copyright 1984 by the American Physical Society]

thermally excited into the band, and the probability of non-radiative transition from the band edge to the DBs is increased. Consequently, the emission intensity at 1.2–1.4 eV tends to decrease at higher temperatures, and it can be considered that the emission efficiency is essentially inversely proportional to the DB density.

With regard to spectral position and form, two models have been proposed. The first of these reflects the density distribution of the band-edge localized states [217, 218], while the second considers the Stokes shift to be based on electron–phonon interactions and phonon–induced broadening [219]. According to the latter, the difference between the excitation spectrum and the emission spectrum, i.e.

the Stokes shift (0.4–0.5 eV) corresponds to the lattice relaxation energy E_r of the electron and/or hole levels. It has been suggested from measurements of the capacitance transient response that the valence-band tail state has a lattice relaxation energy of this magnitude [38], and some experimental data have been reported indicating that the form in the vicinity of the valence band which is involved in the radiative recombination process (optical depth 0.3 ± 0.1 eV) is a self-trapped hole state (termed an A center) [220]. The facts seem to support the latter model, but still fall short in providing definitive evidence.

(b) Transitions involving structural defects

As shown in Fig. 4.35, in doped a-Si:H or non-doped material with high DB densities, photoluminescence over the energy range 0.8–0.9 eV becomes dominant. The origin of this emission is often attributed to a partly radiative transition between the band edge and the DB state [221, 222]. However, optically detected ESR (ODMR) signals suggest that the DB transition is a non-radiative process [220], and it becomes difficult to interpret the photoluminescence at 0.8–0.9 eV in terms of a simple DB transition. On the other hand, there have been reports that the origin of the low-energy photoluminescence is either a DB defect which is charge-coupled with residual nitrogen impurities ($N_2^- - D^-$) in non-doped a-Si:H [220], or the $P_4^+ - D^-$ defect complex (described in Section 4.1) in P-doped a-Si:H [223]. It seems that it will be some time yet before a detailed picture is established.

4.5 Silicon Alloy Materials and Multilayer Film Properties

4.5.1 BASIC PROPERTIES OF SILICON ALLOY MATERIALS

Amorphous silicon alloy materials include narrow-band-gap materials such as a-SiGe and a-SiSn, and wide-band-gap materials such as a-SiC, a-SiN, a-SiO and microcrystalline ($\mu c-$) Si. Owing to space restrictions, it will only be possible to present here a few examples of some typical experimental data, rather than describe the basic properties of these materials in a systematic fashion. It is recommended that the interested reader refers to other handbooks [125, 224] or the original papers for more detailed information on this subject.

Figures 4.36 and 4.37 show the relationships of the conductivities (in the dark and under simulated solar irradiation) to the Tauc gap, in a-SiGe:H (Tauc gap = 1.7–1.0 eV) and a-SiC:H (Tauc gap = 1.8–3.5 eV), respectively [225]. While the film growing process has been optimized in various ways [226], after a full consideration of the various film growing mechanisms, no better film properties have been obtained up until now with these alloy materials than those presented by a-Si, with respect to the photo conductivity or photo conductivity/dark conductivity ratios, for example.

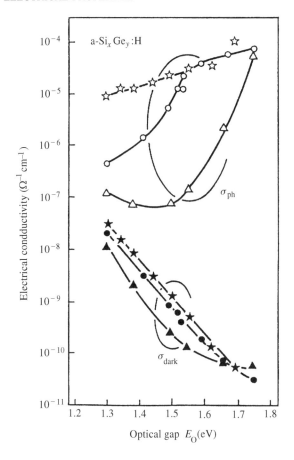

Fig. 4.36 The relationship of photoconductivity (σ_{ph}), and dark conductivity (σ_{dark}) to the optical gap (E_O) in a-SiGe:H, for samples prepared by different plasma CVD techniques: $\triangle$, diode-type; $\circ$, triode-type; $\star$, hydrogen-diluted triode-type [225, reproduced with permission]

While considerable active research effort has been paid to the electrical and optical properties in relation to the structural defects [227], carrier transport properties [228, 229], and localized electron states [230–232], no generalized understanding has been achieved, probably owing to the much more complicated nature of a-SiGe when compared to a-Si. In the case of a-SiC, although adequate data are available on the pn-control by doping and the optical properties [224], reports on the carrier transport, structural defects and localized electron states are somewhat limited. For a-SiN, the conductivity increases when the N content is low (<10 atom%), as the N atoms serve as a dorpant, despite an increased Tauc gap. Moreover, it is known that the Tauc gap increases abruptly from *ca* 3 eV to 5.2 eV or so, at about the stoichiometric composition (Si_3N_4) [233].

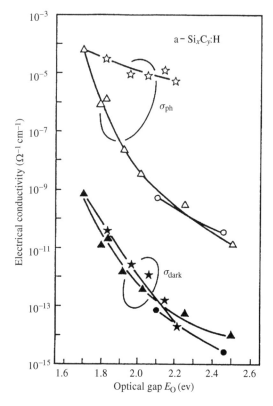

Fig. 4.37 The relationship of photoconductivity (σ_{ph}) and dark conductivity (σ_{dark}) to the optical gap (E_O) in a-SiC:H, for samples prepared by different plasma CVD techniques: △, diode type; ○, triode type; ☆, hydrogen-diluted triode-type [225, reproduced with permission]

4.5.2 PROPERTIES OF MULTILAYER FILMS

(a) Band offset

Most a-Si devices are fabricated in the form of hetero-junction structures, consisting of materials with different band gaps, such as a-Si:H with a-SiC:H or a-SiN:H. The major factors which exert an influence on the electrical properties of such hetero-junctions are band discontinuities, i.e. ΔE_c (conduction band) and ΔE_v (valence band), where $E_{g1} - E_{g2} = \Delta E_c + \Delta E_v < 0$. In the case of amorphous semiconductors, the Tauc gap is generally used for band gap. When using this, it is necessary to take into consideration the fact that the mobility gap, which is of an electrical nature, is larger than the Tauc gap, e.g. by *ca* 0.1–0.2 eV in the case of a-Si:H [234]. The band discontinuities can be characterized by a number of methods, of which the most direct is probably to measure the valence-band spectrum by photoelectron spectroscopy, e.g., X-ray photoemission spectroscopy (XPS). Recently, a method for determining ΔE_c (conduction band) and ΔE_v (valence band) has been developed using the conduction-band spectrum as measured by bremsstrahlung isochromat spectroscopy (BIS) [235].

In addition to independently measuring and comparing the energy positions (with reference to the vacuum level) of the valence-band and conduction-band edges in the two materials constituting the hetero-junction, it is necessary to evaluate the band discontinuities by comparing the spectrum obtained for the actual hetero-junction with the reference spectra of the two component materials, via the spectrum-separation technique. This is because such band discontinuities are not determined by the bulk properties (such as electron affinity) alone, but also strongly depend on the interface charges represented by the electrical dipoles induced by the difference in electrical negativity between the two materials, and the local chemical bonding at the hetero-junction interface [236]. Moreover, it is possible that non-uniformity in the composition, resulting from the preparitive process involved, and other extraneous factors, such as the accumulation or depletion of hydrogen in the interface region, may have an effect.

Figure 4.38 shows the ΔE_v (valence band) data of a-Si:H/a-SiC:H hetero-junctions, obtained by XPS, as a function of the carbon content in a-SiC:H [237]. Since a-SiC:H, as used in solar cells has a Tauc gap of 2.0–2.1 eV (for a

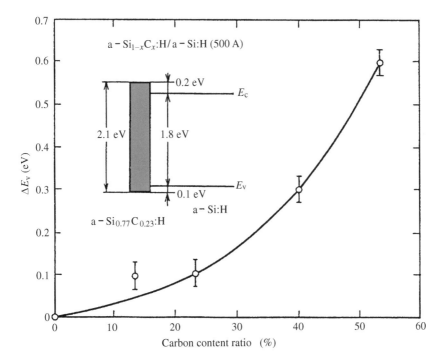

Fig. 4.38 Dependence of the valence-band discontinuity on the carbon content ratio of a-SiC:H in a-SiC:H/a-Si:H hetero-junctions [237, reproduced with permission]

carbon content 10–20 atom%), the valence band has a discontinuity of *ca* 0.1 eV, which may prevent the transport of photo-induced holes in (p) a-SiC:H/i a-Si:H hetero-junction solar cells. From this figure, it can be seen that $\Delta E_v \sim 0.6$ eV and $\Delta E_c \sim 1.1$ eV in stoichiometric a-SiC:H (*ca* 2.8 eV)/a-Si:H.

An interesting hetero-junction band structure has been pointed out for a-Si:H/a-SiN:H, where ΔE_v, determined by the above procedure, has a negative value, i.e. -0.15 eV (a-Si$_{0.62}$N$_{0.38}$:H, 2.2 eV gap). In the case of stoichiometric a-Si$_3$N$_4$:H *ca* 5.2 a-Si:H, the following values have been reported: $\Delta E_v \sim \Delta E_c \sim 1.7$ eV (XPS) [237], and $\Delta E_v \sim 1.2$ eV, $\Delta E_c \sim 2.2$ eV (XPS + BIS) [238]. For a-Si/a-SiGe structures, $\Delta E_v \leq 0.2$ eV seems to hold over a wide range of Ge contents [239]. The band discontinuity has been estimated by the use of internal photoelectric total yield measurements (using a junction) and measurements of the capacity–voltage characteristics, in addition to determinations of the photoelectric total yield. Typical results obtained for a-Si:H/c-Si junctions, for example, are $\Delta E_v = 0.6$–0.7 eV, and $\Delta E_c = 0.05$–0.09 eV [240, 241].

(b) Properties of heterojunctions

In a multilayer film structure built up by the repeated stacking of amorphous semiconductor materials with different compositions, abruptness of in changes of the composition and flatness at the actual hetero-junction become important. Although not discussed further here, it has been demonstrated directly by the analysis of X-ray diffraction patterns and by direct observations using TEM that in some preparative processes the interface can be basically controlled at the atomic-scale level (within a few ångstrons) [242, 243]. The next problem concerns defects at the interface and in the neighboring regions.

It has previously been thought, in the case of amorphous semiconductors without any long-range order, that lattice matching is not required (in contrast to crystalline semiconductors, and various types of heterojunction can be freely formed without producing any interface defects. This argument has been based on the fact that no interface features have been identified in amorphous semiconductors of high bulk defect density (or localized state density), and does not hold any longer for structurally sensitive materials with very low bulk-defect densities, such as a-Si:H. It is expected that disturbances in the chemical bonding, represented by such parameters as the bond length, bond angle and dihedral angle, are inevitably induced in relaxing the stress caused by differences in the chemical bond state, e.g. mean bond length, as well as lattice mismatching. If these effects are sufficiently large, it is therefore no wonder that certain defects, such as dangling bonds, are formed at these sites. Since defect formation requires a relatively extensive atomic rearrangement, this is likely to occur mainly at the time of the hetero-junction formation. Therefore, the quantity of defects formed at an interface depends upon the magnitude of the difference in the chemical structures, and the degree of flexibility of the growing lattice for absorbing the internal stress, plus the nature of the film growing process and the sequence of the hetero-junction formation. Here, the presence of hydrogen, which acts as a stress relaxation agent (or a defect inducing agent), is not to be ignored.

Without giving a lengthy introduction, let us examine some of the experimental data obtained which structures concerns the properties of the interface in amorphous semiconductor hetero-junction. One of the techniques that can be used to investigate the interface level is the electroabsorption (EA) method, described in Section 4.2 [244]. It is possible to determine the internal electric field (or potential) by applying this method to multilayer hetero-junction structures, and also to derive the location and size of the charge and electrical dipoles in the vicinity of the interface through simple reasoning involving electromagnetic theory. Fig. 4.39 shows the dependence of the EA signal intensity on the external biased voltage, as measured in the reflection mode with repeated multilayer films (40 periods) of a-Si:H (20 Å)/a-SiC:H (2.8 eV, 60 Å) [245]. In this figure, V_b is the built-in potential to be determined (for 40 layers of a-Si:H), indicating *ca* 42 mV per a-Si:H layer. The presence of such a built-in potential suggests an asymmetric charge distribution in the repeated multilayer film structure (interfaces for a-Si:H on a-SiC:H and a-SiC:H on a-Si:H). On the basis of detailed data obtained from experiments in which the thickness of each unit layer was

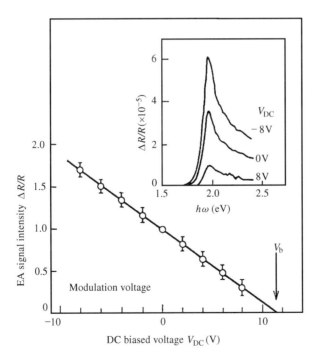

Fig. 4.39 EA spectra measured in the light reflection mode for a multiple-quantum-well structure of a-SiC:H/a-Si:H, and the dependence of the signal intensity on DC biased voltage [245, reproduced with permission. Copyright 1989 by the American Institute of Physics]

varied, it has been found that there are positive charges (2×10^{12} cm^{-2}) at the a-Si:H on a-SiC:H interface and negative charges of the same magnitude at the a-SiC:H on a-Si:H interface. When the same technique was applied to a-Si:H/a-SiN:H and a-Si:H/a-SiO:H structures, it was reported that an electrical dipole of *ca* 10^{12} cm^{-2} exists in the vicinity of the a-Si:H on a-SiN:H interface [246]. These values are to be regarded as representing the minimum limits of the interface defect densities.

Another method for characterizing interfaces based on multilayer film heterojunctions is to utilize light absorption spectra in the lower-energy regions. In this case, the interface defect density is determined by separating the absorption component for the interface from that of the bulk, while changing the number of repeated multilayers. With this technique applied to an a-Si:H/a-SiC:H (*ca* 2.2 eV) multilayer film hetero-structure, a value for the dopole of the order of 10^{10} cm^{-2} has been reported [247], which is significantly smaller than the value found the previous work [246].

Information about the interface levels can also be obtained by measuring the current induced when applying a voltage pulse to a single hetero-structure (e.g. a-SiN:H/a-Si:H), or by analyzing the dependence of the transient current and collected charge on the biased voltage under illumination with light pulses. These techniques have indicated an electron accumulation layer at the a-SiN:H on a-Si:H interface, with an estimated charge density of *ca* 5×10^{11} cm^{-2} [248, 249]. This value is contradictory to the data obtained from EA measurements, but the reason for this discrepancy has not yet been clarified. It should be noted that since the a-SiN:H/a-Si:H hetero-junction is a fundamental structure of the thin film transistor, the dependence of the interface characteristics on the stacking sequence could pose serious problems in designing reliable devices.

(c) Quantum effects

Studies on multilayer film hetero-junction structures in amorphous semiconductors have been motivated by the following factors [243]: (1) to enhance the sensitivity and accuracy of interface characterization by increasing the number of interfaces, (2) to improve composition control and film quality by using multilayer thin films of multicomponent systems which are difficult to realize as simple alloy films, (3) to seek new functions based on the quantum-size or quantum-confinement effects; and (4) to assess those physical parameters which cannot be evaluated from the use of a single layer. Let us now examine some experimental data which suggest the presence of such quantum effects in a-Si:H and investigating their physical significance from the viewpoints of (3) and (4).

First, let us consider the one-dimensional, single-quantum-well structure shown in Fig. 4.40. Assuming that the thickness of the a-Si:H well layer l_w is suitably smaller than the mean free path of the electron l and the inelastic scattering length L_i (phase coherent length), the wavenumber k in the normal direction may be regarded as a good quantum number for describing the electron state,

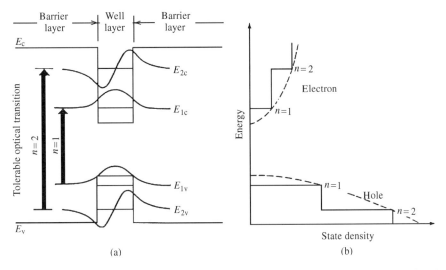

Fig. 4.40 Quantum levels (a) and state-density distributions (b) in a one-dimensional, single-quantum-well structure

with the electron wave becoming a standing wave owing to the interference of multiple reflections in the barrier/well layer/barrier region, thus allowing a wave of a particular wavenumber, k (i.e. quantized). If the barrier is infinitely high, $k = \pi n / l_w$, ($n = 1, 2, 3$, etc), and assuming a parabolic band for the bulk (effective mass m^*), the state density per well layer is given as below by using a two-dimensional state-density function $N_\parallel(E)$ for the non-quantized parallel component:

$$N(E) = \sum_n N_\parallel(E - E_n) \tag{4.114}$$

where:

$$E_n = \frac{1}{2m^*}\left[\frac{\pi\hbar}{l_w}\right]^2 n^2 + E_C \tag{4.115}$$

For example, if the parallel component is regarded as being the crystal parabolic band, then N becomes a step function which rises from 0 to a constant value of $m^*/\pi\hbar^2$ at $E = 0$. The state-density distribution $N(E)$, shown in Fig. 4.40, corresponds to this situation.

Therefore, when the electron wave function is quantized (one-dimensional quantization, in this case), the band edge is raised from E_C to E_1. This effect is reflected in the electrical and optical properties of quantum-well structures in the form of (1) a drop in the parallel conductivity (with the Fermi level held constant), (2) a shift of the absorption spectra into the blue, and (3) a decrease in the carrier capture factor (under weak electron–phonon coupling, multiphonon processes).

In addition, changes may occur in the intensities of the photoluminescence spectra, resulting both from quantum mechanical effects (confinement of the wave function) and semiclassical effects (such as a decrease in the relaxation space available for the excess carriers). In fact, these phenomena have already been observed in various types of a-Si:H quantum-well structures [242, 250, 251]. While these facts suggest the presence of the quantum effect, they fall short of directly demonstrating the various features of the quantized electron structure as shown in Fig. 4.40.

Experimental data that have been recognized as direct evidences for the quantum effect, or for demonstrating quantized structures, include the resonant tunnelling currents observed in double-barrier structures of a-SiN:H/a-SiN:H/a-SiN:H, and the derivative absorption spectra measured for multiple-quantum-well structures of a-Si:H/a-SiC:H. These data are presented in Figs. 4.41 [252] and 4.42 [253], respectively. In Fig. 4.41, weak 'structures' attributable to resonance tunnelling effects can be discerned under the biased voltage conditions corresponding to quantum levels E_n within a well layer. From the results of experiments which involved changing the thickness of the well layer, the effective mass of the electron was estimated to be $m^* = 0.6 \, m_0$, from equation (4.115) with an infinite barrier being assumed.

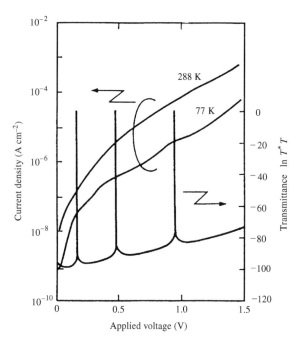

Fig. 4.41 Illustration of the resonance tunnelling current effect in dual-barrier structures of a-SiN:H/a-Si:H (40 Å)/a-Si:H; the transmittance is calculated by using the Wentzel–Kramers–Brillouin (WKB) approximation [252, reproduced with permission. Copyright 1987 by the American Physical Society]

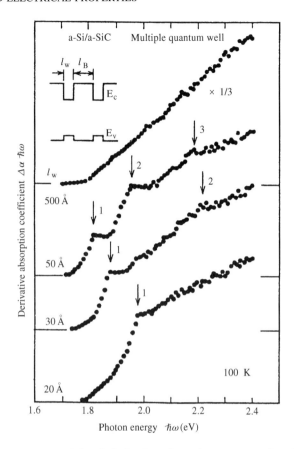

Fig. 4.42 Temperature-modulated derivative absorption spectra of multiple-quantum-well structures in a-Si:H/a-SiC:H; arrows indicate the thresholds of the tolerable optical transitions shown in Fig. 4.40(a), i.e. energy intervals between the quantum levels in the conduction and valence bands [253, reproduced with permission. Copyright 1988 by the American Physical Society and American Institute of Physics]

Figure 4.42 requires some additional explanation. In the crystalline quantum-well structure, the inter-band optical transition occurs while maintaining the quantum number n and wavenumber in the parallel direction k. If the exciton effect is ignored, the absorbance spectrum takes a 'staircase' form, reflecting the state-density distribution shown in Fig. 4.40. In the case of the amorphous semiconductor, according to a simple non-direct transition model (the Tauc model) which neglects the selection rule for the parallel wavenumber k, it seems that an approximate form of the absorption spectrum can be expressed by a convolution integral of the valence band- and conduction-band sub-band state densities for the same quantum number n. Therefore, an absorption spectrum of a segmented form, with kinks at each of the sub-band transition energy positions may be anticipated. Such a weak kinked structure may be obscured by certain disorder effects which are specific to amorphous materials, so that it cannot actually be observed in the absorption spectrum, except for an overall shift towards the blue. In Fig. 4.42, the absorbance spectrum is differentiated in order to turn such a kink into the readily observable staircase form (a derivative spectrum based on temperature modulated spectroscopy). An interpretation can be made that the behavior of the spectral form and the quantum structure position on changing the well layer width is based on the quantum effect. On the basis of experimental data obtained by applying similar derivative measurements to other quantum-well systems such as a-Si:H/a-SiO$_2$, the reduced masses of the electron and the hole are estimated to be 0.23 m_0.

While the effective mass values estimated from the electrical and optical data involve some discrepancies, the matter of key importance is that the thickness of an a-Si:H well layer is less than 50 Å and that structures attributable to quantum effects can be identified. This means that the inelastic scattering length L_i is not that small in comparison to 50 Å at the measurement temperatures (100 K or lower, in general). This fact does not contradict the results obtained from theoretical studies of the pre-exponential factor for the electrical conductivity (described in Section 4.3). The next point to be considered is the magnitude of the mean free path l determined by the structural disorder which does not affect the coherence of the electron wave function. If only a one-dimensional degree of freedom in the normal direction is taken into account, the state density takes the form of a delta-function-sequence structure, which has a value at each energy point E_n in the case of an infinitely large mean free path, or in the absence of disorder. If disorder is introduced, the delta-function-like state density changes into a broader distribution with a finite energy width (ΔE). If the blur in the energy grows larger than the intervals between the neighbouring quantum levels, it is found that structures based on quantum effects become difficult to observe experimentally. Let us examine this situation by applying some simple reasoning. On the basis of the uncertainty principle, the lifetime of the quantized state is given by $\hbar/\Delta E$, and the mean free path l corresponds to the expression $(\hbar/\Delta E)\sqrt{2E_n/m^*}$, (with $E_c = 0$, being assumed). This corresponds to the case where a damping term proportional to the velocity is used, and the state-density distribution takes the form of a function of the Lorentzian type, with FWHM $=$ $\Delta E/2$. The value of the FWHM is given by the following expression:

$$\Delta E_n/2 = \frac{\hbar}{l}\sqrt{E_n/2m^*} \qquad (4.116)$$

With the conditions for observing the above-mentioned structure being taken into consideration, and using the following equation:

$$E_{n+1} - E_n > (\Delta E_{n+1} + \Delta E_n)/2 \qquad (4.117)$$

it can be concluded that $l > l_w/\pi$. By considering the fact that structures based on quantum effects are experimentally visible when the well layer width l_w is smaller than 50 Å, the mean free path in a-Si:H must be ca 16 Å, or even larger. It has been shown theoretically that the experimental trends shown in Fig. 4.42 can be reproduced if the mean free path or degree of disturbance is of this order [254].

At least, for the case of an electron at the conduction-band edge, this value is not contradictory with that suggested by interpretation of the optical and electrical properties described previously in Sections 4.2 and 4.3. It has been pointed out that if the mean free path is of this size (i.e. not of the order of the interatomic distance), the effective mass m^*, calculated by applying equation (4.114) to the experimental data, is almost equal to the value for the reference crystalline system with zero disorder, as long as it is allowed to establish its existence [254]. If this suggestion is correct, it could be regarded as providing important ideas for examining the electron structure of a-Si:H by comparison with crystalline Si. While few reports are currently available which are based on this viewpoint, it can confidently be expected that observations of quantum effects will lead to new findings regarding the electron structure, and the optical and electrical properties of amorphous semiconductors.

References

1. D. Weaire and M. F. Thorpe, *Phys. Rev.*, **B4** (1971) 2508.
2. F. Yonezawa and M. H. Cohen, *Fundamental Physics of Amorphous Semiconductors*, edited by F. Yonezawa, Springer-Verlag, (1981) 119.
3. D. Adler, *Hydrogenated Amorphous Silicon, Part A: Semiconductors and Semimetals*, edited by J. I. Pankove, Academic Press, **21** (1984) 291.
4. W. B. Jackson, S. M. Kelso, C. C. Tsai, J. W. Allen and S. -J. Oh, *Phys. Rev.* **B31**, (1985) 5187.
5. F. Wooten, K. Winer and D. Weaire, *Phys. Rev. Lett.* **54** (1985) 1392.
6. R. Biswas, C. Z. Wang, C. T. Chan, K. M. Ho and C. M. Soukoulis, private communications (1990).
7. J. D. Joannoupoulos and M. L. Cohen, *Phys. Rev.* **B7** (1977) 2644.
8. T. Hama and F. Yonezawa, *Solid State Commun.* **29** (1979) 371.
9. T. M. Hayes, J. W. Allen, J. L. Beeby and S.-J. Oh, *Solid State Commun.* **56** (1985) 953.
10. B. von Roedern, L. Ley and M. Cardona, *Phys. Rev. Lett.*, **39** (1977) 1576.
11. D. A. Papaconstantopoulos and E. N. Economou, *Phys. Rev.*, **B24** (1981) 7233.
12. D. P. Divincenzo, J. Bernholc and M. H. Brodsky, *Phys. Rev.*, **B28** (1983) 3246.
13. A. D. Zdetsis, E. N. Economou, D. A. Papaconstantopoulos and N. Flytzanis, *Phys. Rev.*, **B31** (1985) 2410.
14. W. B. Jackson and S. B. Zhang, *Advances in Disodered Semiconductors*, edited by H. Fritzsche, World Scientific, Singapore, **3**, (1990) 63; W. B. Jackson, C. C. Tsai and C. Doland, *Phil. Mag.*, **B64** (1991) 611.
15. E. N. Economou, C. M. Soukoulis, M. H. Cohen and S. John, *Disordered Semiconductors*, edited by M. A. Kastner, Plenum Press, New York (1987) 681.
16. B. I. Halperin and M. Lax, *Phys. Rev.*, **148** (1966) 722.
17. W. Sritrakool and V. Sa-yakanit, *Phys. Rev.*, **B33** (1986) 1199.
18. S. John, C. Soukoulis, M. H. Cohen and E. N. Economou, *Phys. Rev. Lett.*, **57** (1986) 1777.
19. P. W. Anderson, *Phys. Rev.*, **109** (1958) 1492.
20. A. Kawabata, *Frontiers of Physics, No. 13*, edited by Y. Otsuki, Kyoritsu Publishing Company, Tokyo (1986) 69 (in Japanese).

21. S. F. Edwards, *New Developments in Semiconductors*, edited by P. R. Wallace, Noordhoff International Publishing, Amsterdam, (1973) 249.
22. E. N. Economou, C. M. Soukoulis, M. H. Cohen and A. D. Zdetsis, *Phys. Rev.*, **B31** (1985) 6172.
23. R. A. Street, J. Zesch and M. J. Thompson, *Appl. Phys. Lett.*, **43** (1983) 672; R. A. Street, *Appl. Phys. Lett.*, **41** (1982) 1060.
24. R. A. Street and N. F. Mott, *Phys. Rev. Lett.*, **35** (1975) 1293.
25. D. K. Biegelsen and M. Stutzmann, *Phys. Rev.*, **B33** (1986) 3006.
26. M. Stutzmann and D. K. Biegelsen, *Phys. Rev. Lett.*, **60**, (1988) 1682.
27. N. Ishii and T. Shimizu, *Phys. Rev.*, **B42** (1990) 9697.
28. K. Morigaki, *Advances in Disodered Semiconductors*, edited by H. Fritzsche, World Scientific, Singapore, **1–A**, (1989) 595.
29. S. R. Elliott, *Phil. Mag.*, **B38** (1978) 325.
30. D. Adler, *Phys. Rev. Lett.*, **41** (1978) 1755.
31. Y. Bar-Yam and J. D. Joannopoulos, *Phys. Rev. Lett.*, **56** (1986) 2203.
32. P. W. Anderson, *Phys. Rev. Lett.*, **34** (1975) 953.
33. H. Okamoto and Y. Hamakawa, *Solid State Commun.*, **24** (1977) 1197.
34. D. Adler and E. J. Yoffa, *Phys. Rev. Lett.*, **36** (1976) 1197.
35. For example, M. Stutzmann, W. B. Jackson, R. A. Street and D. K. Biegelesen, *Disordered Semiconductors*, edited by M. A. Kastner, Plenum Press, New York, (1987) 407.
36. J. Tauc and Z. Vardeny, *Phil. Mag.*, **52** (1985) 313.
37. Z. Vardeny, T. X. Zhou, H. Stoddart and J. Tauc, *Solid State Commun.*, **65** (1988) 1049.
38. A. V. Gelatos, J. D. Cohen and J. P. Harbison, *Appl. Phys. Lett.*, **49** (1986) 722.
39. K. Hattori, H. Okamoto and Y. Hamakawa, *Phil. Mag.*, **B57** (1988) 13.
40. J. E. Northrup, *Phys. Rev.*, **B40** (1989) 5875 .
41. R. A. Street and D. K. Biegelsen, *J. Non-Cryst. Solids*, **35/36** (1980) 651.
42. T. Shimizu, H. Kidoh, X. Xu, A. Morimoto and M. Kumeda, *Matter. Res. Symp. Proc.*, **118** (1988) 665.
43. J. Ristein, J. Hautala and R. C. Taylor, *Phys. Rev.*, **B40** (1989) 88.
44. S. Yamasaki, H. Ohkushi, A. Matsuda, K. Tanaka and J. Isoya, *Phys. Rev. Lett.*, **65** (1990) 756.
45. J. D. Cohen, J. P. Harbison and K. W. Wecht, *Phys. Rev. Lett.*, **48** (1982) 109.
46. J. M. Essick and J. D. Cohen, *Phys. Rev. Lett.*, **64** (1990) 3062.
47. H. M. Branz and M. Silver, *Phys. Rev.*, **B42** (1990) 7420.
48. G. Schumm and G. H. Bauer, *Phil. Mag.*, **B64** (1991) 515 .
49. K. Morigaki and M. Yoshida, *Phil. Mag.*, **B52** (1985) 289.
50. D. V. Lang, J. D. Cohen and J. P. Harbison, *Phys. Rev.*, **B25** (1982) 5285.
51. H. Ohkushi, *Phil. Mag.*, **B52** (1985) 33.
52. H. Ohkushi, T. Takehara, Y. Tokumaru, Y. Yamasaki, H. Oheda and K. Tanaka, *Phys. Rev.*, **B27** (1983) 5184.
53. J. D. Cohen and D. V. Lang, *Tetrahedrally Bonded Amorphous Semiconductors*, edited by D. Adler, Plenum Press, New York (1985) 299.
54. H. Kida, K. Hattori, H. Okamoto and Y. Hamakawa, *J. Appl. Phys.*, **59** (1986) 4079.
55. H. Oheda, *J. Appl. Phys.*, **52** (1981) 6693.
56. K. Abe, H. Okamoto and Y. Hamakawa, *Phil. Mag.*, **B58** (1988) 171.
57. K. Tanaka, H. Ohkushi and S. Yamasaki, *Tetrahedrally Bonded Amorphous Semiconductors*, edited by D. Adler, Plenum Press, New York (1985) 239.
58. J. Kocka, M. Vanecek and A. Triska, *Advances in Disodered Semiconductors*, edited by H. Fritzsche, World Scientific, Singapore, **1–A** (1989) 298.
59. K. Winer, *Phys. Rev. Lett.*, **63** (1989) 1487.

60. G. D. Cody, C. R. Wronski, B. Abeles, R. B. Stephens and B. Brooks, *Solar Cells*, **2** (1980) 227.
61. S. Yamasaki, *Phil. Mag.*, **B56** (1987) 79.
62. W. B. Jackson, N. M. Amer, A. C. Boccara and D. Fournier, *Appl. Opt.*, **20** (1981) 1333.
63. J. Kocka, M. Vanecek and F. Schauer, *J. Non-Cryst. Solids*, **97/98** (1987) 715.
64. H. Okamoto and Y. Hamakawa, *J. Non-Cryst. Solids*, **77/78** (1985) 1441.
65. J. Tauc, *Amorphous and Liquid Semiconductors*, edited by J. Tauc Plenum Press, New York (1974) Ch. 4, 159.
66. K. Tanaka, *J. Appl. Phys.*, **20** (1981) 267.
67. M. H. Cohen, C. M. Soukoulis and E. N. Economou, *AIP Conf. Proc.*, **120** (1984) 371; M. H. Cohen, C. M. Soukoulis and E. N. Economou, *J. Non-Cryst. Solids*, **77/78** (1985) 171.
68. G. D. Cody, *Hydrogenated Amorphous Silicon, Part B: Semiconductors and Semimetals*, edited by J. I. Pankove, Academic Press, New York, **21** (1984) ch. 2, 11: G. D. Cody, *Physics and Applications of Amorphous Semiconductors, No. 2*, edited by F. Demichelis, World Scientific, Singapore (1988) 28.
69. W. E. Pickett, D. A. Papaconstantopoulos and E. N. Economou, *Phys. Rev.* **B28** (1983) 2232.
70. A. Frova and A. Selloni, *Tetrahedrally Bonded Amorphous Semiconductors*, edited by D. Adler, Plenum Press, New York, (1985) 271.
71. S. Abe and Y. Toyozawa, *J. Phys. Soc. Jpn*, **50** (1981) 2185.
72. F. Yonezawa and F. Sato, *J. Phys. Soc. Jpn*, **57** (1988) 1797.
73. H. Okamoto, K. Hattori and Y. Hamakawa, *J. Non-Cryst. Solids*, **137/138** (1991) 627; H. Okamoto, K. Hattori and T. Hamakawa, *J. Non-Cryst. Solids*, **198–200** (1996) 124.
74. E. N. Economou, *Physics and Applications of Amorphous Semiconductors, No. 1*, edited by F. Demichelis, World Scientific, Singapore (1987) 3.
75. G. D. Cody, T. Tiedje, B. Abeles, B. Brooks and Y. Goldstein, *Phys. Rev. Lett.*, **47** (1981) 1480.
76. S. Grisp and L. Ley, *J. Non-Cryst. Solids*, **59/60** (1983) 253; K. Winer and L. Ley, *Advances in Disodered Semiconductors*, edited by H. Fritzsche, World Scientific, Singapore, **1–A** (1989) 365.
77. S. Aljishi, J. D. Cohen, S. Jin and L. Ley, *Phys. Rev. Lett.*, **64** (1990) 2811.
78. For example, see *Modulation Technique : Semiconductors and Semimetals*, Vol. 9, edited by R. K. Willardson Academic Press, New York, (1972).
79. For example, Y. Hamakawa, *Hydrogenated Amorphous Silicon, Part B: Semiconductors and Semimetals*, Vol. 21, edited by J. I. Pankove, Academic Press, New York, **21** (1984) ch. 5, 141.
80. G. Weiser, U. Dersch and P. Thomas, *Phil. Mag.*, **B57** (1988) 721.
81. W. B. Jackson and N. M. Amer, *Phys. Rev.*, **B25** (1982) 5559 .
82. A. V. Gelatos, K. K. Mahavadi and J. D. Cohen, *Appl. Phys. Lett.*, **53** (1988) 403.
83. I. Hirabayashi, *Kotai Buturi*, **20** (1985) 27 (in Japanese); I. Hirabayashi and K. Morigaki, *Phil. Mag.*, **54** (1986) L119.
84. P. O'Connor and J. Tauc, *Phys. Rev.*, **B25** (1982) 2748.
85. G. Lucovsky, *Solid State Commun.*, **3** (1965) 299.
86. M. Jaros, *Deep Levels in Semiconductors*, Adam Hilger, (1982) 156.
87. H. Okamoto, H. Kida, T. Kamada and Y. Hamakawa, *Phil. Mag.*, **B52** (1985) 1115.
88. H. Ohkushi and K. Tanaka, *Phil. Mag. Lett.*, **55** (1987) 135.
89. S. Yamasaki, S. Kuroda and K. Tanaka, *AIP Conf. Proc.*, **157** (1987) 9.
90. E. A. Abrahams, P. W. Anderson, D. C. Licciardello and T. W. Ramarkrishnan, *Phys. Rev. Lett.*, **42** (1979) 685.

91. H. Fukuyama, *Frontiers of Physics, 2*, edited by Y. Otsuki, Kyoritsu Publishing Company, Tokyo (1984) 59 (in Japanese).
92. A. Kawabata, *Solid State Commun.*, **38** (1981) 823; A. Kawabata, *J. Phys. Soc. Jpn*, **54** (1984) 318.
93. N. F. Mott, *Phil. Mag.*, **B49** (1984) L75; N. F. Mott, *Phil. Mag.*, **B51** (1985) 19; N. F. Mott, *Phil. Mag.*, **B52** (1985) 177; N. F. Mott, *Phil. Mag.*, **B58** (1988) 369. Mott has given corrections to equation 4.45 which are applicable to the region of the band edge.
94. E. N. Economou, M. H. Cohen and C. M. Soukoulis, *J. Non-Cryst. Solids*, **77/78** (1985) 151; C. M. Soukoulis, M. H. Cohen, E. N. Economou and A. D. Zdetsis; *J. Non-Cryst. Solids*, **77/78** (1985) 47.
95. N. F. Mott, *Conduction in Non-Crystalline Materials*, Oxford Science Publishing, Oxford (1987) ch. 3.
96. N. F. Mott and E. A. Davis, *Electronic Processes in Non-Crystalline Materials*, Clarendon Press, Oxford, 1979.
97. Y. Imry, *Phys. Rev. Lett.*, **44** (1980) 469.
98. H. Overhof and P. Thomas, *Electronic Transport in Hydrogenated Amorphous Silicon*, Springer Tracts in Modern Physics, Springer-Verlag, Berlin, **114** (1989) Ch. 5.
99. H. Müller and P. Thomas, *J. Phys. (Paris)*, **C17** (1984) 5337; H. Müller and P. Thomas, *J. Phys. (Paris)*, **C18** (1985) 3191; H. Müller and P. Thomas, *J. Phys. (Paris)*, **C18** (1985) 5815.
100. W. E. Spear and C. S. Cloude, *Phil. Mag.*, **B58** (1988) 467; P.G. LeComber and W. E. Spear, *Phys. Rev. Lett.*, **25** (1970) 509.
101. For example, see M. Pollak, *Non-Crystalline Semiconductors*, Volume I, edited by M. Pollak, CRC Press, Boca Raton (1987) Ch. 5B.
102. N. F. Mott, *J. Non-Crystalline Solids*, **1** (1968) 1.
103. M. L. Knotek, *Solid State Commun.*, **17** (1975) 1431.
104. H. Okamoto and Y. Hamakawa, *J. Non-Crystalline Solids*, **33** (1979) 225.
105. H. Overhof, *Disordered Semiconductors*, edited by M. A. Kastner, Plenum Press, New York (1987) 713.
106. H. Fritzche, *Solar Cells*, **3** (1980) 447 .
107. P. N. Butcher and L. Friedman, *J. Phys. (Paris)*, **C42** (1977) 3803; P. N. Butcher and L. Friedman, *Phil. Mag.*, **B50** (1984) L5.
108. D. Emin, *Solid State Commun.*, **22** (1977) 409, D. Emin, *Phys. Rev.*, **B30** (1984) 5766; D. Emin, *Phil. Mag.*, **B51** (1985) L53.
109. H. Overhof and W. Beyer, *Phil. Mag.*, **B43** (1981) 433; H. Overhof and W. Beyer, *Phil. Mag.*, **B47** (1983) 377.
110. R. A. Street, *Phys. Rev. Lett.*, **49** (1982) 1187 .
111. B. E. Springett, *J. Appl. Phys.*, **44** (1973) 2925.
112. J. Stuke, *J. Non-Cryst. Solids*, **97/98** (1987) 1.
113. H. Overhof, *J. Non-Cryst. Solids*, **97/98** (1987) 539.
114. J. Kakalios and R. A. Street, *Phys. Rev.*, **B34** (1986) 6014; J. Kakalios and R. A. Street, *J. Non-Cryst. Solids*, **97/98** (1987) 769.
115. W. B. Jackson, C. C. Tsai and S. M. Kelso, *J. Non-Cryst. Solids*, **77/78** (1985) 218.
116. P. G. Lecomber, D. I. Jones and W. E. Spear, *Phil. Mag.*, **35** (1977) 1173.
117. L. Friedman, *J. Non-Cryst. Solids*, **6** (1971) 329; L. Friedman, *Phil. Mag.*, **38** (1978) 467.
118. D. Emin, *Phil. Mag.*, **35** (1977) 1189 .
119. L. Friedman, *Non-Crystalline Semiconductors*, Volume I, edited by M. Pollak, CRC Press, Boca Raton (1987) Ch. 5A.
120. Refined arguments are given in H. Okamoto, K. Hattori and Y. Hamakawa, *J. Non-Cryst. Solids*, **164–166** (1994) 445.

121. N. F. Mott, *Phil. Mag.*, **B63** (1991) 3.
122. W. E. Spear, G. Willeke, P. G. LeComber and A. G. Fitzgerald, *J. Phys. (Paris)*, **42** (1981) C4-252.
123. W. E. Spear and P. G. LeComber, *Phil. Mag.*, **33** (1976) 935.
124. P. G. LeComber and W. E. Spear *Amorphous Semiconductors*, edited by M. H. Brodsky, Topics in Applied Physics, Springer-Verlag, Berlin, **36**, (1979) Ch. 9.
125. H. Okamoto, *Comprehensive Technology for New Materials and Processes*, edited by R. Yamamoto, *et al.* (1987) Section 6(2), 453 (in Japanese) .
126. R. A. Street and J. Zesch, *Phil. Mag.*, **B50** (1984) L19.
127. R. A. Street, D. K. Biegelsen, W. B. Jackson, N. M. Johnson and M. Stutzmann, *Phil. Mag.*, **B52** (1985) 235.
128. M. Stutzmann, *Phil. Mag.*, **B53** (1986) L15 .
129. M. Stutzmann, D. K. Biegelsen and R. A. Street, *Phys. Rev.*, **B35** (1987) 5666.
130. F. J. Kampas and P. E. Vanier, *Phys. Rev.*, **B31** (1985) 3654.
131. K. Winer and W. B. Jackson, *Phys. Rev.*, **B40** (1989) 12558; K. Winer, R. A. Street, N. M. Johnson and J. Walker, *Phys. Rev.*, **B42** (1990) 3120.
132. R. A. Street, *Phys. Rev. Lett.*, **49** (1982) 1187.
133. Y. Bar-Yam, D. Adler and J. D. Joannopoulos, *Phys. Rev. Lett.*, **57** (1986) 467.
134. H. Okamoto and Y. Hamakawa, *J. Non-Cryst. Solids*, **33** (1979) 225.
135. R. A. Street and K. Winer, *Phys. Rev.*, **B40** (1989) 6236 .
136. S. Zufar and E. A. Street, *Phys. Rev.*, **B40** (1989) 5235.
137. K. Pierz, W. Fuhs and H. Mell, *Phil. Mag.*, **B63** (1991) 123.
138. G. Muller, *Appl. Phys.*, **A45** (1988) 41.
139. M. Stutzmann and J. Stuke, *J. Non-Cryst. Solids*, **66** (1984) 145; M. Stutzmann and R. A. Street, *Phys. Rev. Lett.*, **54** (1985) 1836.
140. I. Hirabayashi, K. Morigaki, S. Yamasaki and K. Tanaka, *AIP Conf. Proc.* **120** (1984) 8.
141. N. Ishii, M. Kumeda and T. Shimizu, *Solid State Commun.*, **53** (1985) 543.
142. S. Yamasaki, S. Kuroda, H. Ohkushi and K. Tanaka, *J. Non-Cryst. Solids*, **77/78** (1985) 339.
143. D. V. Lang, J. D. Cohen and J. P. Harbison, *Phys. Rev. Lett.*, **48** (1982) 421.
144. R. A. Street and J. Kakalios, *Phil. Mag.*, **B54** (1986) L21; R. A. Street, J. Kakalios, C. C. Tsai and T. M. Hayes, *Phys. Rev.*, **B35** (1987) 1316; R. A. Street, M. Hack and W. B. Jackson, *Phys. Rev.*, **B37** (1988) 4209.
145. R. Banerjee, T. Furui, H. Ohkushi and K. Tanaka, *Appl. Phys. Lett.*, **53** (1988) 1829.
146. W. B. Jackson, *Phys. Rev.*, **B41** (1990) 12323 .
147. E. Z. Liu and W. E. Spear, *Phil. Mag.*, **B64** (1991) 245.
148. G. Krotz, J. Wind, H. Stitzl, G. Muller, S. Kalbitzer and H. Mannsperger : Phil. Mag. B63 (1991) 101.
149. R. A. Street : Phil. Mag. B49 (1984) L15 .
150. H. Scher, *Photoconductivity and Related Phenomema*, edited by J. Mort and D. M. Pai, Elsevier, (1976) Ch. 3.
151. D. Mencaraglia and J. P. Kleider, *Phil. Mag. Lett.*, **55** (1987) 63.
152. See, for example: B. K. Ridley, *Quantum Processes in Semiconductors*, Oxford Science Publications, Oxford (1988) Ch. 6, J. Bourgoin and M. Lannoo, *Point Defects in Semiconductors, II*, Springer-Verlag, Berlin (1983) Ch. 4, Ch. 6, M. Jaros, *Deep Levels in Semiconductors*, Adam Hilger, (1982) Ch.8.
153. K. Huang and A. Phys, *Proc. R. Soc.*, (London) **A204** (1950) 406.
154. C. H. Henry and D. V. Lang, *Phys. Rev.*, **B15** (1977) 989.
155. B. K. Ridley, *J. Phys. (Paris)*, **C13** (1980) 2015.
156. N. Robertson and L. Friedman, *Phil. Mag.* **B33** (1978) 753.

157. K. Hattori, Y. Niwano, H. Okamoto and Y. Hamakawa, *J. Non-Cryst. Solids*, **137/138** (1991) 363.

158. A. Doghmana and W. E. Spear, *Phil. Mag.*, **B53** (1986) 463.

159. M. Lax, *Phys. Rev.*, **119** (1965) 1502.

160. W. E. Spear and P. G. Le Comber, *Phil. Mag.*, **B52** (1985) 247.

161. J. Tauc, *Hydrogenated Amorphous Silicon, Part B: Semiconductors and Semimetals*, Vol. 21, edited by J. I. Pankove, Academic Press, New York (1984) Ch. 9, A. Mourchid, R. Vanderhaghen, D. Hulin and P. M. Fauchet, *Phys. Rev.*, **B42** (1990) 7667.

162. L. Onsager, *Phys. Rev.*, **54** (1938) 554.

163. D. M. Pai and R. C. Enck, *Phys. Rev.*, **B11** (1975) 5163.

164. H. Okamoto, T. Yamaguchi and Y. Hamakawa, *J. Phys. Soc. Jpn*, **49** (1980) Suppl. A, 1213.

165. F. Carasco and W. E. Spear, *Phil. Mag.*, **B47** (1983) 495.

166. W .A. Schiff, R. I. Delvin, H. T. Grahn, J. Tauc and S. Guha, *Appl. Phys. Lett.*, **54** (1989) 1911.

167. H. Okamoto, H. Kida and Y. Hamakawa, *Phil. Mag.*, **B49** (1984) 231.

168. W. Schockley and W. T. Read, *Phys. Rev.*, **87** (1952) 835.

169. J. G. Simmons and G. W. Taylor, *Phys. Rev.*, **B4** (1971) 502.

170. V. Halperun, *Phil. Mag.*, **B54** (1986) 473.

171. W. E. Spear and P. G. Le Comber, *Photoconductivity and Related Phenomema*, edited by J. Mort and D. M. Pai, Elsevier, Amsterdam (1976) Ch. 6.

172. C. R. Wronski and R. E. Daniel, *Phys. Rev.*, **B23** (1981) 794.

173. S. Oda, Y. Saitoh, I. Shimizu and E. Inoue, *Phil. Mag.*, **B43** (1981) 1079.

174. T. Kagawa, N. Matsumoto and K. Kumabe, *Phys. Rev.*, **B28** (1983) 4570.

175. F. Villant and D. Jousse, *Phys. Rev.*, **B34** (1986) 4088; F. Villant, D. Jousse and J.-C. Bruyere, *Phil. Mag.*, **B57** (1988) 649.

176. E. Morgado, *Phil. Mag.*, **B63** (1991) 529.

177. M. Hack, S. Guha and M. Shur, *Phys. Rev.*, **B30** (1984) 6991; M. Hack, S. Guha and M. Shur, *Appl. Phys. Lett.*, **45** (1984) 467.

178. H. Okamoto, *Amorphous Semiconductor Technologies and Devices*, edited by Y. Hamakawa, JARECT, Volume 16, Ohm and North-Holland, Amsterdam (1984) Ch. 4.1.

179. M. A. Kastner and D. Monroe, *Solar Energy Mat.*, **8** (1982) 41.

180. A. Rose, *Concepts in Photoconductivity and Allied Problems*, Wiley-interscience, London (1963), Ch. 3.

181. J. Hubin, E. Sauvain and A. V. Shah, *IEEE Trans.* **ED36** (1989) 2789.

182. K. Hattori, H. Okamoto and Y. Hamakawa, *Phys. Rev.*, **B45** (1992) 1126.

183. F. W. Schmidlin, *Phys. Rev.*, **B16** (1977) 2362 .

184. J. Noolandi, *Phys. Rev.* **B16** (1977) 4466.

185. J. Orenstein, M. A. Kastner and V. Vaninov, *Phil. Mag.*, **46** (1982) 23; J. Orenstein and M. A. Kastner, *Phys. Rev. Lett.*, **46** (1981) 1421.

186. T. Tiedje and A. Rose, *Solid State. Commun.*, **37** (1980) 49; T. Tiedje, J. M. Cebulka, D. L. Morel and B. Abels, *Phys. Rev. Lett.*, **46** (1981) 1425; T. Tiedje, *Appl. Phys. Lett.*, **40** (1982) 627; T. Tiedje, *Hydrogenated Amorphous Silicon, Part C: Semiconductors and Semimetals*, edited by J. I. Pankove, Academic Press, New York, (1984) 207.

187. M. Marshall, R. A. Street, M. J. Thompson and W. B. Jackson, *Phil. Mag.*, **B57** (1988) 387, plus references contained therein.

188. Y. Tsutsumi, K. Uchiyama, K. Hattori, H. Okamoto and Y. Hamakawa, *Technical Digest PVSEC-5*, Kyoto (1990) 809.

189. A. C. Hourd and W. E. Spear, *Phil. Mag.*, **B51** (1985) L13.

190. J. Marshall, R. A. Street and M. J. Thompson, *Phil. Mag.*, **B54** (1986) 51.

191. D. Goldie and W. E. Spear, *Phil. Mag.*, **B57** (1988) 135.

192. C. Cloude, W. E. Spear, P. G. Le Comber and A. C. Hourd, *Phil. Mag.*, **B54** (1986) L113.
193. R. E. Johanson, Y. Kaneko and H. Fritzsche, *Phil. Mag.*, **B63** (1991) 57.
194. D. Monroe, *Phys. Rev. Lett.*, **54** (1985) 54 .
195. W. E. Spear and C. S. Cloude, *Phil. Mag.*, **B58** (1988) 467.
196. M. Kemp and M. Silver, *Phil. Mag.*, **B63** (1991) 437.
197. W. E. Spear and H. Steemers, *Phil. Mag.*, **B47** (1983) L77.
198. E. Trammell and F. J. Walter, *Nucl. Instrum. Methods*, **76** (1969) 317.
199. E. A. Schiff, *Phil. Mag. Lett.*, **55** (1987) 87.
200. J. Kocka, C. E. Nebel and C. D. Abel, *Phil. Mag.*, **B63** (1991) 221.
201. D. Ritter, E. Zeldove and K. Weiser, *Appl. Phys. Lett.*, **49** (1986) 791; D. Ritter, E. Zeldove and K. Weiser, *J. Appl. Phys.* **62** (1987) 4563; D. Ritter, E. Zeldove and K. Weiser, *Phys. Rev.*, **B38** (1988) 8296; D. Ritter, E. Zeldove and K. Weiser, *Advances in Disodered Semiconductors*, edited by H. Fritzsche, World Scientific, Singapore, **1B** (1989) 871.
202. A. R. Moore, *J. Appl. Phys.*, **54** (1983) 222; A. R. Moore, *Hydrogenated Amorphous Silicon, Part C: Semiconductors and Semimetals*, Vol. 21, edited by J. I. Pankove, Academic Press, New York, (1984) 239.
203. J. C. Heuvel, M. J. Geerts and J. W. Metselaar, *J. Appl. Phys.*, **68** (1990) 1281.
204. H. Okamoto, H. Kida, S. Nonomura and Y. Hamakawa, *J. Appl. Phys.*, **54** (1983) 3236.
205. B. Faughnan, A. Moore and R. Crandall, *Appl. Phys. Lett.*, **44** (1984) 613.
206. W. E. Spear and P. G. Le Comber, *Phil. Mag.*, **B52** (1985) 247.
207. H. Fritzsche and K.-J. Chen, *Phys. Rev.*, **B28** (1983) 4900.
208. R. A. Street, J. Kakalios and M. Hack, *Phys. Rev.*, **B38** (1988) 5603.
209. H. Overhof and M. Silver, *Phys. Rev.*, **B39** (1989) 10426.
210. W. E. Spear and H. L. Steemers, *Phil. Mag.*, **B47** (1983) L107; J. Kakalios, *Phil. Mag. Lett.*, **55** (1987) 129; D. Golodie, P. G. Le Comber and W. E. Spear, *Phil. Mag.*, **B58** (1988) 107.
211. See, for example: M. Silver, E. Snow, B. Wright, M. Aiga, L. Moore, V. Cannella, R. Ross, S. Payson, M. P. Shaw and D. Adler, *Phil. Mag.*, **B47** (1983) L39; M. Silver, D. Adler, M. P. Shaw and V. C. Cannella, *Phil. Mag.*, **B53** (1986) L89; M. Silver, G. Winborne, D. Adler and V. Cannella, *Appl. Phys.*, **50** (1987) 983; D. Adler and M. Silver, *Phil. Mag. Lett.*, **56** (1987) 113.
212. W. E. Spear, H. L. Steemers and H. Mannsperger, *Phil. Mag.*, **B48** (1983) L49.
213. W. E. Spear, H. L. Steemers, P. G. Le Comber and R. A. Gibson, *Phil. Mag.*, **B50** (1984) L33.
214. J. Bullot, P. Cordier, M. Gauthier and G. Mawawa, *Phil. Mag.*, **B55** (1987) 599.
215. R. A. Street, D. K. Biegelsen and R. L. Weisfield, *Phys. Rev.*, **B30** (1984) 5861.
216. R. A. Street, *Phil. Mag.*, **B37** (1978) 35.
217. D. J. Dunstan and F. Boulitrop, *J. Phys. (Paris)*, **42** (1981) Suppl. 10, C4-331.
218. T. M. Searle and W. A. Jackson, *Phil. Mag.*, **B60** (1989) 237.
219. R. A. Street, *Hydrogenated Amorphous Silicon, Part B: Semiconductors and Semimetals*, Vol. 21, edited by J. I. Pankove, Academic Press, New York (1984) 197.
220. K. Morigaki, *Kotai Buturi*, **20** (1985) 528 (in Japanese); K. Morigaki, *Advances in Disodered Semiconductors*, edited by H. Fritzsche, World Scientific, Singapore **1–A** (1989) 595.
221. R. A. Street, *Phys. Rev.*, **B21** (1980) 5775.
222. B. A. Wilson, A. M. Sergent, K. W. Wecht, A. J. Williams, T. P. Kerwin, C. M. Taylor and J. P. Harbison, *Phys. Rev.*, **B30** (1984) 3320.
223. M. Tajima, H. Ohkushi, S. Yamasaki and K. Tanaka, *Phys. Rev.*, **B33** (1986) 8522.
224. See, for example: *Latest Handbook for Amorphous Silicon*, edited by K. Takahashi and M. Konagi, Science Forum, (1983) (in Japanese); *Amorphous Semiconductor*

Technologies and Devices, edited by Y. Hamakawa, JARECT, Volumes 16 and 22, Ohmsa and North-Holland, Pub. Co. Tokyo and Amsterdam (1984, 1987).

225. K. Tanaka, *Optoelectronics*, 4 (1989) 143.
226. A. Matsuda, M. Koyama, N. Ikeuchi, Y. Imamura and K. Tanaka, *Jpn J. Appl. Phys.*, **25** (1986) L54.
227. M. Stutzmann, R. A. Street, C. C. Tsai, J. B. Boyce and S. E. Ready, *J. Appl. Phys.*, **66** (1989) 569.
228. R. A. Street, C. C. Tsai, M. Stutzumann and J. Kakalios, *Phil. Mag.*, **B56** (1987) 289.
229. F. Karg, W. Kruhler, M. Moller and K. v. Klitzing, *J. Appl. Phys.*, **60** (1986) 2016.
230. H. Matsumura, *J. Appl. Phys.*, **64** (1988) 1964.
231. Y. Tsutsumi, H. Okamoto and Y. Hamakawa, *Phil. Mag.*, **B60** (1989) 695.
232. L. Chen, J. Tauc, J.-K. Lee and E. Schiff, *Phys. Rev.*, **B43** (1991) 11694.
233. M. Hirose, *Jpn. J. Appl. Phys.*, **21** (1981) Suppl. 1, 275.
234. C. R. Wronski, S. Lee, M. Hicks and S. Kumer, *Phys. Rev. Lett.*, **63** (1989) 1420.
235. T. Takahashi, *Kotai Buturi*, **23**, (1988) 397 (in Japanese).
236. See, for example: *Heterojunction Band Discontinuities*, edited by F. Capasso and G. Margaritondo, North-Holland, Amsterdam (1987); *Band Structure Engineering in Semiconductor Microstructures*, edited by R. A. Abram and M. Jaros, Plenum Press, New York, (1988), plus references contained therein.
237. M. Hirose, *Technical Digests, PVSEC-3*, Tokyo (1987) 651.
238. W. B. Jackson, *Extended Abstracts of the 18th Conference on Solid State Devices and Materials*, Tokyo (1986) 691.
239. F. Evangelist, S. Modesti and F. Boscherini, *MRS Symp. Proc.*, **49** (1985) 95.
240. H. Mimura and Y. Hatanaka, *Appl. Phys. Lett.*, **50** (1987) 326.
241. J. M. Essick and J. D. Cohen, *Appl. Phys. Lett.*, **55** (1989) 1232.
242. B. Abels and T. Tiedje, *Hydrogenated Amorphous Silicon, Part C: Semiconductors and Semimetals*, edited by J. I. Pankove, Academic Press, New York, **21** (1984) 407.
243. For a special issue devoted to amorphous semiconductors, multilayer structures and superlattices see *Phil. Mag.*, **B60** (1989).
244. S. Nonomura, H. Okamoto and Y. Hamakawa, *Appl. Phys.*, **A32** (1983) 31.
245. K. Hattori, M. Tsujishita, H. Okamoto and Y. Hamakawa, *Appl. Phys. Lett.*, **55** (1989) 763; K. Hattori, M. Tsujishita, H. Okamoto and Y. Hamakawa, *Optoelectronics*, **4** (1989) 155.
246. C. B. Roxlo, B. Abeles and T. Tiedje, *Phys. Rev. Lett.*, **52** (1984) 1994; C. B. Roxlo, B. Abeles and T. Tiedje, *Phys. Rev.*, **B34** (1986) 2522.
247. A. Asano, T. Ichimura, M. Ohsawa, H. Sakai and Y. Uchida, *J. Non-Cryst. Solids*, **97/98** (1987) 971.
248. R. A. Street, M. J. Thompson and J. M. Johnson, *Phil. Mag.*, **B51** (1985) 1; R. A. Street, M. J. Thompson and J. M. Johnson, *AIP Conf. Proc.*, **120** (1984) 410.
249. R. A. Street and C. C. Tsai, *Appl. Phys. Lett.*, **48** (1986) 1672.
250. M. Hirose and S. Miyazaki, *J. Non-Cryst. Solids*, **97/98** (1987) 23.
251. K. Hattori, T. Mori, H. Okamoto and Y. Hamakawa, *Appl. Phys. Lett.*, **51** (1987) 1259.
252. S. Miyazaki, Y. Ihara and M. Hirose, *Phys. Rev. Lett.*, **59** (1987) 125.
253. K. Hattori, T. Mori, H. Okamoto and Y. Hamakawa, *Phys. Rev. Lett.*, **60** (1988) 825; K. Hattori, T. Mori, H. Okamoto and Y. Hamakawa, *Appl. Phys. Lett.*, **53** (1988) 2170.
254. K. Hattori, H. Okamoto and Y. Hamakawa, *Physics and Applications of Amorphous Semiconductors*, edited by F. Demichelis, World Scientific, Singapore (1988) 92; K. Hattori, H. Okamoto and Y. Hamakawa, *J. Non-Cryst. Solids*, **114** (1989) 687.

5 STRUCTURAL STABILITY AND PHOTO-INDUCED EFFECTS

This Chapter will be concerned with structural metastability, which is an essential feature of the amorphous solid. First, the various photo-induced effects that have been observed in amorphous semiconductors such as the chalcogenides will be overviewed and their characteristic features will be described in order. This will be followed by a general description of the similarities and dissimilarities between the photo-induced metastable defects in both crystalline and amorphous semiconductors, pointing out the structural factors that are related to the generation of metastable defects in amorphous silicon. In this connection, details of the Staebler–Wronski effect will be presented, together with a discussion of the models that have been proposed, the underlying physics, and some of the problems that have been encountered. With reference to the thermal stability, the diffusion of hydrogen and crystallization of multilayer films will also be discussed.

5.1 Nature of the Non-Equilibrium Solid

In Section 1.1.2, it was stated that the amorphous solid is a non-equilibrium substance which takes different macroscopic states depending upon the particular preparative process that has been used. As shown in Fig. 1.2 (curve D), the free energy G of a system does not take a minimum value. For this reason, an external disturbance, either thermal or optical, will induce structural changes at the macro- or microscopic levels to increase or decrease the free energy of the system, either reversibly or irreversibly. This is an essential feature of amorphous semiconductor materials, on account of their non-equilibrium solid nature. This issue is of extreme importance, as it is directly related to the structural stability from the applications point of view.

While Fig. 1.2 shows the general macroscopic state of the system, it must be represented in a different form if we wish to consider the relaxation of the amorphous structure at the atomic level. It can be seen that each of the atoms in the amorphous structure has multiple points of metastability, with the overall

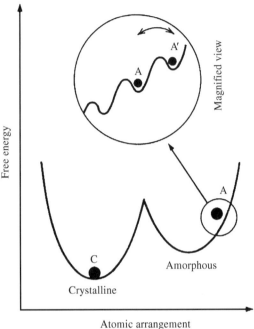

Fig. 5.1 Illustration of the free-energy changes in amorphous solids by using configurational coordinates: A $\longrightarrow$ C represents crystallization (macroscopic), and A $\longleftrightarrow$ A$'$ the photo-induced effect and its reverse process (microscopic)

structure being relaxed through the various transition over these points one after the other. For our discussions in this present chapter, the representation of free energy by using configurational coordinates, as depicted in the enlarged inset of Fig. 5.1, may be more helpful. The transition from the amorphous state (A) to the crystalline state (C) along the free-energy curve shown in this figure is a first-order phase transition involving crystallization, which is a macroscopic structural change (as described in Chapter 1). The crystallization of a-Si:H will be described in Section 5.4.1. In Fig. 5.1, the microscopic transition from A to A$'$ represents a photo-induced effect, which is the central theme of this present chapter, i.e. a localized structural change induced by light while keeping the amorphous structure unchanged. In most cases, this change can be eliminated by annealing at temperatures lower than the glass transition temperature, with the reversibility of the A$'$ $\rightleftharpoons$ A process being maintained.

These reversible photo-induced effects owe much to the network flexibility. In particular, according to Phillips' definition of the mean coordination number M, such photo-induced effects are likely to be manifested most markedly in amorphous solids with $2 < M < 3$, which can change into 'glassy' materials (see Fig. 2.2 in Chapter 2). However, it should be noted that some photo-induced phenomena such as those described in the next section have been observed in a-Si:H with $M \sim 3.5$. It seems that hydrogen plays an important role in these phenomena, as has been discussed in detail in Chapters 2 and 3.

In this way, crystallization caused by the non-equilibrium nature of the system and the reversible photo-induced effects both relate to the essential properties of these systems. They are not only of great significance from the point of view of applications, but also provide great academic interest.

5.2 Photo-Induced Effects in Amorphous Semiconductors

The photo-induced reversible change or metastability shown by A $\rightleftharpoons$ A′ in Fig. 5.1 has been observed in various forms; these can be classified as shown in Table 5.1, which presents different aspects depending upon the category of the material. There are two major types: (1) the metastability which can be characterized by increased structural disorder (photostructural change) and a red-shift of the photo-absorption edge (photodarkening), and (2) that which can be characterized by an increased density of the paramagnetic centers (ESR spin density) and a broad absorption band below the absorption-edge energy.

5.2.1 PHOTODARKENING AND PHOTOSTRUCTURAL CHANGES

Since these phenomenon are observed only in the chalcogenide semiconductors and not in the corresponding crystals, they may be regarded as 'amorphous-specific phenomena' [1]. In connection with this form of metastability, the following changes have been recognized: the randomness in the structure increases on illumination (photostructural change), the ESR spin density does not increase, the absorption edge shifts toward the lower-energy region (photodarkening), and the original state is restored only if the sample is annealed just below the glass transition temperature (T_g) [1]. If the sample is illuminated at low temperatures, the ESR spin density increases at the same time, but this can be attributed to the superposition of a further phenomenon which will be described later [2]. In both a-As and a-Si:H, neither photodarkening or photostructural changes are found to occur.

5.2.2 PHOTO-INDUCED METASTABLE ESR CENTERS

In chalcogenide glasses, a varieties of photo-induced ESR centers have been observed in addition to photodarkening [3]. In such phenomena, broad absorption bands in the energy region below the optical absorption edge (optical gap E_0) are created [3]. In the case of a-As$_2$S$_3$, both kinetically fast and slow processes take place, which display different excitation spectra ($h\nu_e$) and annealing temperatures (T_{ann}) [4], as shown in Table 5.1. Both processes are more unstable than those that occur in photodarkening, and the metastable ESR centers that result from either process disappear at room temperature [4].

These findings provide an experimental basis for the defect model of the chalcogenide glasses (D^+–D^- pair, or C_3^+–C_1^- pair), and support Anderson's

Table 5.1 Photo-induced reversible changes (metastability) in amorphous semiconductors

Phenomenon	Photodarkening and photostructural changes in chalcogenides[a]	Photo-induced metastable ESR (paramagnetic) centers[a,b]			
		Chalcogenide (e.g. As_2S_3)		a-Si:H (non-doped)	
Photoexcitation energy hv_e	$hv_e \lesseqgtr E_0$	Fast process	Slow process	$g = 2.0055$ (SW)[c]	$g = 2.004$, (2.011) (LESR)[d]
Annealing temperature T_a	$T_a \sim T_g^e$	~180 K	~300 K	~450 K	~60 K
Optical changes	Absorption edge shift $\Delta E_0 < 0$	Absorption band below photo-absorption edge $\Delta\alpha(hv < E_0) > 0$			
Structural changes	Quantitative randomness increased, no bonding defects formed	Formation of $C_1^0(S)$, $P_2^0(As)$, $P_1^0(As)$ ($U_{eff} < 0$)		$T_3^0 (U_{eff}?)$; distance T_3–H is greater than 4.2 Å	Wave function similar to T_3^0; carriers at weak-bond states or of other regions?
		As (close to an sp hybridzed orbital)	As (close to a p orbital); homo-bond (As–As, S–S)		
Specific to Amorphous state?	Yes	Undecided			
Specific to material?	Specific to chalcogenides	Common to all amorphous semiconductors			

[a] E_O, optical absorption-edge energy (optical gap)

[b] C, chalcogen atom; P, pnictide atom; T, tetrahedral atom; U_{eff}, effective electron correlation energy

[c] SW, Staebler–Wronski effect (see Section 5.3)

[d] LESR, light-induced ESR (see Section 3.7.3)

[e] Glass transition temperature

proposal of a negative effective electron correlation energy, U_{eff} [5−7]. The origin of a negative U_{eff} is attributed to the presence of lone-pair electrons in the chalcogen atom (S or Se), with their energies being higher than that of the As−S (or As−Se) bonding orbitals and lying at the top of the valence band. For this reason, lone-pair electrons display varied behaviors depending upon the structure of the surrounding network, such as forming bonds with neighbouring atoms or moving electrons into other non-bonding orbitals by breaking covalent bonds. This situation, together with the low value of the mean coordination number, contributes to an extensive network flexibility.

The origin of the metastable ESR center is attributed to a hole being captured by a 3p orbital of the S atom (C_1^0), or an electron being trapped by a non-bonding orbital of the twofold coordinated As atom (P_2^0: pnictide) on the basis of analysis of the hyperfine interactions with the nuclear spin of As [8, 9]. The difference between the fast and slow processes is explained in the following way: in the induced twofold coordinated As centers (P_2^0), the former center is close to an sp hybridized orbital, while the latter is close to a 4p orbital, in relation to the As−As homo-bonding [8, 9].

On the other hand, no systematic experimental data are available for the photo-induced ESR of chalcogenide crystals such as As_2S_3 or As_2Se_3. In experiments which involved irradiating a sample with X-rays at 77 K, the induction of paramagnetic (ESR active) centers was not observed in either c-As_2S_3 or c-As_2Se_3 [10]. Nevertheless, it seems premature to generally conclude that this phenomenon never occurs in crystalline materials (see Table 5.1).

In a-Si:H, three kinds of photo-induced ESR centers have been observed. As shown in Table 5.1, the corresponding g-values for the three centers are 2.0055, 2.004 and 2.011. Phenomenologically, these centers can be split up into two categories, i.e. one involving metastable ESR centers which are not removed after thermal annealing' at room temperature ($g = 2.0055$) and the other which consists of centers which can only exist at low temperatures ($g = 2.004$ and 2.011).

The former group corresponds to the Staebler−Wronski effect where photoconductivity and dark conductivity are reduced after illumination with light for long periods of time. In this group, changes in the various properties are attributed to the creation of metastable ESR centers (defects) with g-values which are equal to those of dangling bonds [11, 12] (see Section 3.5.2). The centers are stable at room temperature, but disappear after annealing at 450 K [11]. A detailed description of the Staebler−Wronski effect will be given in Section 5.3.

According to preliminary interpretations, the two ESR centers in the latter group involve electrons which are captured at the conduction-band-tail states ($g = 2.004$) or holes which are captured at the valence-band-tail states ($g = 2.011$). These centers are observed under illumination at temperatures lower than 150 K [11], but disappear immediately if the illumination is turned off. At temperatures lower than 50 K, the centers disappear slowly, even after illumination [13]. In parallel with these ESR observations, a broad absorption band is seen to extend towards the lower-energy side in the absorption spectrum below the optical gap. Analysis of the shoulder in the ESR spectrum, which is based on hyperfine interactions with the ^{29}Si nuclear spins, suggests that the centers for $g = 2.004$ and 2.011 resemble dangling bond [13]. However, more detailed analysis is required before any definite conclusions can be made about this. While not shown in Table 5.1, these metastable ESR center are also observed in a-As, and can be regarded as being common to all of the amorphous semiconductors [14].

5.2.3 STRUCTURAL ORIGINS OF PHOTO-INDUCED METASTABILITY

Let us now discuss in some detail the significant features of these photo-induced phenomena from the viewpoints of structural flexibility and the electron–lattice interactions involved in the photo excitation processes.

Photodarkening and photostructural changes are only observed in chalcogenide semiconductors, and not in a-As and a-Si:H. As stated above, one of the most important factors with respect to structural flexibility is the mean coordination number M of the amorphous semiconductor material (see Section 2.1.1). It has been definitely confirmed that photo-induced changes occur when $2 < M < 3$, and not when $3 < M < 4$. On the other hand, in the case of photo-induced metastable ESR centers, these conditions are somewhat relaxed so that they can occur at $2 < M < 3.5$ (e.g. in a-Si:H, $M \sim 3.5$). This phenomenon does not require as much structural flexibility in order to change the network randomness as is needed in the case of photostructural change, because it basically concerns localized defect formation within the network, and does not require a change in the structural order of the entire network. In fact, the photo-induced effects of metastable defects have been observed in crystals of both III–V and II–VI compound semiconductors, as well as I–VII alkali halides [15, 16].

With respect to the mean coordination number, most of the III–V and II–VI compound semiconductors have a fourfold coordination ($M = 4$) structure which is based on sp^3 hybridized orbitals. Nevertheless, photo-induced defects occur owing to the ionic nature of the bond, which is another important factor concerning the structural stability. According to Phillips' definition, ionicity is the ratio of the force constant for bond stretching to that of bond bending, and a bond with a higher ionicity is more flexible in bond bending [17]. Antisite As in GaAs crystals and color centers in alkali halides are examples of photo-induced metastable defects. A silicon crystal, which has the same fourfold coordination structure based on sp^3 hybridized orbitals, the so-called tetrahedrally bonded structure, has a typically rigid structure owing to the lack of ionicity of the bonds, and is devoid of any photo-induced metastable defects. Recently, it has been reported that implanting hydrogen or boron into silicon crystals generates metastable defects [18]. It has been theoretically suggested that hydrogen has two points of stability, namely a bond-center site and a tetrahedral site [19]. However, this metastability is based on extrinsic defects caused by the H or B themselves, and should be distinguished from the intrinsic defects present in the regular lattice of Si.

An electron and a hole excited by the band-gap illumination may be trapped or recombined at a defect, or interact with the defect as an exciton. If the electron–lattice interaction is strong, it can be expected that a metastable state will be readily induced. Toyozawa and coworkers, in systematic theoretical studies [20, 21] have advocated the electron–phonon coupling constant g_{ep}, defined below, as an important physical parameter for discussing the metastability of crystals at deep levels:

$$g_{ep} = E_{LR}/B \qquad (5.1)$$

where E_{LR} is the lattice relaxation energy, either optical or acoustic in type, and B is the kinetic energy of the fully localized electron, represented by $B = (h^2/2m^*)(\pi/a)^2$, where the wave function is Gaussian in nature, h is the Planck constant, m^* is the effective mass of the electron, and a is the lattice constant. Without involving detailed discussions, it may be claimed that the larger g_{ep} is, then the more readily a metastable localized level can be formed.

If crystalline Si is viewed from this standpoint, it may be expected that g_{ep} is not so large as to form metastable localized levels, based on the grounds that there is a lack of interaction with optical phonons (absence of ionicity in the bonding) and the small value of m^*. With regard to electron–lattice interactions, the localized nature of the defect level is important for inducing local lattice deformation. Since the energy gap of c-Si is as small as 1.1 eV, it is rather difficult to form deeper levels. As is known empirically, the Bohr radius r_B of the wavefunction for the defect level is related to the bound energy of the defect $|E_\infty - E|$ in the following way [22]:

$$r_B \propto 1/|E_\infty - E|^{1/2} \qquad (5.2)$$

where in case of an electron, E_∞ corresponds to the bottom of the conduction band, E_C. This means that the deeper the defect level energy is, then the greater is the degree of localization of the electron state. For example, when the defect level is 0.1 or 0.5 eV, r_B is ca. 10 or 5 Å, respectively [22]. If the empirical formula shown in equation (5.2) is applied, it may be claimed that in c-Si of a smaller energy gap, the degree of localization for the electron state of the defect level cannot become large enough [2, 23]. If a wavefunction is spatially extended, it may be difficult to expect an electron–lattice interaction to induce locally a large lattice deformation [23].

In view of the importance of electron localization, it may be said that structural randomness is a factor in inducing localization by its very nature, in comparison to the regular crystalline lattice.

To sum up the above discussion, the formation of metastable defects may involve the following factors: (1) a mean coordination number M which meets the condition $2 < M < 3.5$, and ionicity of the bond (from the viewpoint of structural flexibility), (2) a greater lattice relaxation energy E_{LR}, i.e. a large ionicity of the bond and a greater effective mass of the electron, m^* (from the viewpoint of electron–lattice interactions), and (3) a greater optical energy gap and structural randomness (from the viewpoint of localization of the defect state).

On comparing a-Si:H with c-Si, while considering these factors, it may be claimed that a-Si:H satisfies every requirement for creating metastable defects because (1) $M \cong 3.5$ (for c-Si, $M = 4$), (2) there is a certain degree of ionicity in the bonds owing to the presence of H, (3) the optical energy gap is close to 1.8 eV (for c-Si, 1.1 eV), and (4) it has a disordered structure.

5.3 The Staebler–Wronski Effect

In 1977, Staebler and Wronski found that both the dark and photo-conductivities of a-Si:H, after being illuminated with band-gap light for a long period of time were reduced, with restoration being achieved by annealing at high temperatures [11]. This reversible photo-induced effect is named the Staebler–Wronski effect, after its discoverers. While many researchers have made great efforts to fully identify this phenomenon, no decisive conclusions have yet been reached as to its mechanism (based on various microscopic models, as described below), nor have any effective measures been found to prevent its occurrence.

5.3.1 DESCRIPTION OF THE PHENOMENON

Figure 5.2 shows the reversible photo-induced changes in the dark and light conductivities that were first observed by Staebler and Wronski [11]. Following illumination, the level of the dark conductivity falls by a few orders of magnitude from state A to state B, while that of the photoconductivity is also reduced by about one order of magnitude. The reduced photoconductivity after illumination (state B) persists for a long time at room temperature, thus reducing the conversion efficiency in any potential photovoltaic applications. For this reason, the phenomenon is also known as photodegradation. If the temperature is raised, the relaxation process is activated and both the dark conductivity and photoconductivity levels are fully restored to their pre-illumination values (state A).

The level of photoconductivity is decided by the generation efficiency of the electron–hole pairs, η, the drift mobility, μ_d, and the carrier lifetime τ. It is known, that among these factors, the lifetime of the carrier (or electron) is considerably reduced by the effect of illumination [24]. On the other hand, the decrease in dark conductivity is mainly caused by an increase in its thermal activation energy, E_σ, thus suggesting a shift of the Fermi level toward the gap center [11].

In this connection, the photo-induced effect may be witnessed by changes in other properties. In photoluminescence (PL), the emission efficiency of the main band (1.3–1.4 eV) is reduced after illumination, while the efficiency of the sub-band (0.8–0.9 eV), attributable to defects, rises [25].

These effects suggest that the illumination induces new localized levels in the energy gap, which is directly supported by the experimental evidence of the photo-induced ESR centers shown in Table 5.1. Since the first ESR measurements on these materials by Dersch *et al.*, who reported an increase in the spin density from 10^{15}–10^{16} cm^{-3} (state A) to 10^{16}–10^{17} cm^{-3} (state B) [12], a number of

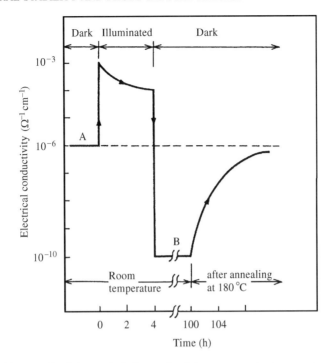

Fig. 5.2 Photoconductivity and dark conductivity measured at room temperature of a-Si:H, reduced by illumination and restored by annealing at 180°C (the Staebler–Wronski effect); illumination by 200 mW cm^{-2} white light (600–900 nm)

studies have been carried out and much progress has been made in elucidating this phenomenon [26, 27].

Figure 5.3 shows ESR spectra of a-Si:H both before and after illumination [27]. The spectra obtained by measuring the ENDOR (electron nuclear double resonance) signal involving the hydrogen nucleus (^{1}H) as a function of the magnetic field, indicate hyperfine structures for the interactions between the ^{29}Si nuclear spins and the electron spins as distinct shoulders (compare with Fig. 3.18). As is evident from this figure, the ESR centers which have been augmented by the illumination are essentially the same as those which existed under dark conditions, i.e. dangling bonds. As described earlier in Section 3.5.2, it is suggested that DBs existing in the dark state are isolated in space from one another. This means that the metastable DBs induced by the illumination are also isolated spatially. The illumination creates DBs of $g = 2.005$, which disappear on annealing at temperatures higher than 160°C, thus causing the Staebler–Wronski effect.

The increase in DBs induced by the illumination leads to an additional absorption band in the photon-energy region below the optical-absorption edge [28]. Fig. 5.4 shows optical absorption spectra measured by photoacoustic spectroscopy (PAS), both before and after illumination [28]. As previously

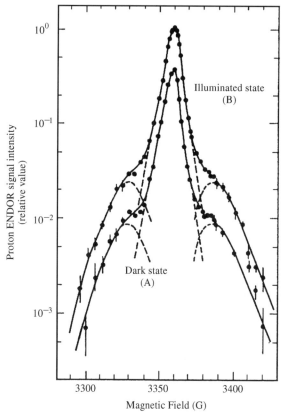

Fig. 5.3 ESR spectra before and after illumination as measured by using proton ENDOR spectroscopy; the spin density in A (dark state) is 4×10^{15} cm^{-3} [27, reproduced with permission. Copyright 1987 by the American Institute of Physics]

mentioned in connection with Table 5.1, it is expected that localized levels are formed in an energy range which is considerably deeper than the band-edge region.

On the basis of these arguments, the following conclusions may be reached:

(1) A metastable DB (D^0) acts as a recombination center for the photo-excited carrier, thus reducing the carrier lifetime τ and the photoconductivity.

(2) A metastable DB serves as both a quenching center for the PL main band, and as an emission center for the PL sub-band.

(3) The presence of the metastable DB causes a change in the energy distribution of the density of the localized states, shifting E_F toward the mid-gap and increasing the activation energy E_σ for dark conductivity.

In this way, the Staebler–Wronski effect can be understood in a global sense.

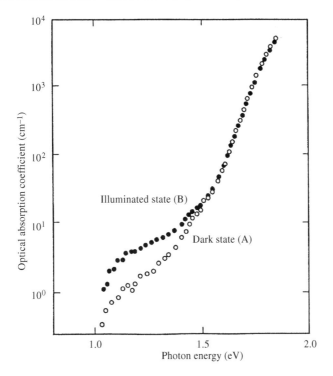

Fig. 5.4 Optical-absorption-edge spectra of a-Si:H before and after illumination as measured by photoacoustic spectroscopy (PAS) (Reproduced by permission of Dr S. Yamasaki, Electrotechnical Laboratory)

It has been confirmed, with regard to the dependence of the Staebler–Wronski effect on doping, that the dark conductivity increases following the illumination, but only with 'barely n-type' a-Si:H with an activation energy for dark conductivity, E_σ, in the range between 0.9 and 1.0 eV [29]. In other n-type and p-type regions, the dark conductivity drops as in the case of non-doped (n-type) samples. For n-type a-Si:H doped with P, detailed experiments involving ESR (^{31}P) hyperfine structure studies and isothermal capacitance transient spectroscopy (ICTS) have been reported [27, 30]. There will be discussed further in Section 5.3.3.

5.3.2 ASPECTS OF KINETICS

As shown in Table 5.1, the wavelength of light needed to induce the Staebler–Wronski effect does not always require photon energies higher than the bandgap [2]. However, photo-induction shows the best efficiency for wavelengths around the band gap, i.e., optical absorption edge of the larger absorption coefficient or light-absorption edge. A similar effect can be achieved by carrier injection, which is important for identifying the mechanism involved [31].

It is thought phenomenologically, therefore, that the metastable dangling bonds which cause the Staebler–Wronski effect are induced either by the direct and indirect non-radiative recombination of photo-induced electron–hole pairs, or by the trapping of one of the two carriers. As discussed in Section 5.2.3, a typical mechanism of energy transfer which leads to photo-induced metastable defects is thought to be the non-radiative recombination of excited electron–hole pairs through strong electron–lattice coupling. It seems reasonable to assume that this recombination proceeds via a strongly localized state. We will now discuss here the macroscopic time sequences involved in the photo-induced creation of metastable DBs and in the annihilation of them by annealing.

Stutzmann measured the photo-induced ESR spin density $N_s (\equiv N_d$, dangling bond density) as a function of the intensity and time t of illumination, and obtained the results shown in Fig. 5.5 [32]. In order to explain this effect, he proposed a model in which metastable DBs are formed as a consequence of the non-radiative recombination of photo-excited electron–hole pairs via the conduction-band-tail and valence-band-tail states.

The increment of defect density N_d is given by the following:

$$dN_d/dt = C_{SW}A_t n\, p \tag{5.3}$$

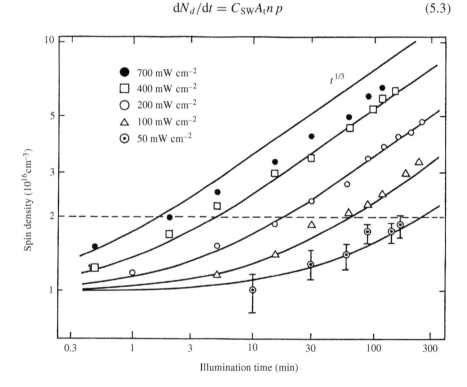

Fig. 5.5 Increase in spin density (dangling bond density) as a function of illumination time using illumination intensity as a parameter [32, reproduced with permission. Copyright 1985 by the American Physical Society]

$$\propto C_{SW}G^2/N_d^2, \tag{5.4}$$

where A_t is the rate constant for the non-radiative recombination between the tail states, C_{SW} is the proportion of such non-radiative recombinations (per unit time) involved in creating defects, n and p represent the concentrations of free electrons and holes, respectively, and G is the generation rate of the electron–hole pairs in proportion to the illumination intensity. In order to derive equation (5.4), it is assumed that the density of the electrons in the conduction-band-tail states and that of the holes in the valence-band-tail states are proportional to n and p, respectively, and that the lifetime of the electron–hole recombination event in the tail states is inversely proportional to N_s [32].

From equation (5.4), the following relationship can be derived:

$$N_d(t)^3 - N_d(0)^3 \propto C_{SW}G^2 t \tag{5.5}$$

which may be represented, if $C_{SW}G^2 t >> N_d(0)^3$, as follows:

$$N_d(t) \propto C_{SW}^{1/3} G^{2/3} t^{1/3} \tag{5.6}$$

This relationship agrees well with the experimental data shown in Fig. 5.5.

Hack et al. demonstrated that the same explanation is valid if the recombination of electrons occurs at a specific center located between the valence-band edge and hole-trapping quasi-Fermi level [33]. In this case, if the hole trap is located in the middle of the mobility gap, then N_d depends upon $[Gt]^{1/2}$ [33].

The process of extinguishing defects (or metastable DBs) created by the illumination (duration t_{ill}), i.e. $N_{met} \equiv N_d(t_{ill}) - N_d(0)$, through annealing at a higher temperature, may be regarded as a thermal relaxation process which is peculiar to amorphous materials. The decrease in N_{met} (at a fixed temperature) with time cannot be represented by a simple exponential function, thus suggesting that the activation energy is not of a single value, but is distributed around 1 eV, within a range of ± 0.2 to 0.3 eV [32]. The relationship can be expressed in a more generalized form, as a stretched exponential decay function, as follows [34]:

$$N_{met}(t) = N_{met}(0) \exp \left[-\left(\frac{t}{\tau} \right)^\beta \right] \tag{5.7}$$

where $N_{met}(0) (\equiv N_d(t_{ill}) - N_d(0))$ is the density of the metastable DBs produced under illumination, t is the duration of annealing at a fixed temperature, τ is a decay constant, and β is a parameter with a value between 0.8 and 0.9. It is believed that the recovery process at higher temperatures is affected extensively by the diffusion of hydrogen [34].

In order to take into consideration the thermal recovery process, the following mathematical 'model' for the defect formation process is proposed in place of equations (5.3) and (5.4) [35]:

$$\frac{dN_d}{dt} = Bt^{-\alpha}[g - \gamma(N_d - N_{d0})] \tag{5.8}$$

and

$$N_{d\infty} - N_d = (N_{d\infty} - N_{d0}) \exp(-Ct^{1-\alpha}) \qquad (5.9)$$

where g is the creation rate of the defects, γ the annihilation rate by thermal annealing, N_{d0} the density of DBs in the dark, $N_{d\infty}$ the density of total existing metastable DB sites plus N_{d0} (approximately equal to the saturated density of the DBs under illumination), α the dispersion parameter ($0 < \alpha < 1$), and B and C are constants. These latter expressions provide a more precise simulation of the system, including the saturation of N_d, and also reproduce the stretched exponential decay process of the annealing [36]. This model requires further detailed studies in order to achieve a detailed evaluation of its applicability.

5.3.3 STRUCTURAL ORIGINS AND MECHANISMS OF DEFECT FORMATION: MICROSCOPIC MODELS

As described in Sections 3.5.2, 5.2.3 and 5.3.1, the defects that are created by the effects of illumination are those with paramagnetic centers of $g = 2.005$, which are equivalent to the native defects that exist in a-Si:H in the dark, i.e. dangling bonds (DBs, D^0). The result of experimental studies on the hyperfine structure of photo-created DBs with ^{29}Si nuclear spin has revealed that they are essentially identical to DBs in the dark, as shown in Fig. 5.3.

It can be claimed that the DBs produced by illumination have the same structural environment as those existing in the dark. In other words, the requirements to be met in the mechanism of photo-induced defect (DB) formation are as follows: (1) spacings between the DBs must be greater than 10 Å [27], and (2) as indicated by experiments carried out using samples of a-Si:H in which the hydrogen has been replaced by deuterium (D), no hydrogen exists in the immediate neighbourhood of the DB [37].

Models for the structural defects and DB-producing mechanisms that have been proposed so far may be largely divided into two main categories: (1) models based on intrinsic structural defects involving Si and H alone, and (2) models based on extrinsic structural defects involving impurities such as oxygen, carbon and nitrogen [38].

Those models which are based on intrinsic structural defects may be further subdivided into a number of different types [38], as shown by the two-dimensional diagrams in Fig. 5.6. In model (a), two DBs are formed by the simple dissociation of weak Si−Si bonds (weak bonds (WBs), see Fig. 3.16 in Chapter 3), while model (b) shows the pairing of a DB with a floating bond (FB, see Fig. 3.16) [39]. In model (c), a charged DB pair is converted into two metastable neutral DBs under the assumption of negative effective electron correlation energies [40], while in model (d), a weak Si−Si bond with a neighbouring Si−H bond is 'switched' to leave two stabilized DBs [38]. In contrast to models (a)−(d), model (e) in Fig. 5.6 is an example of an extrinsic-structural-defect model, where X represents an element from group V and a

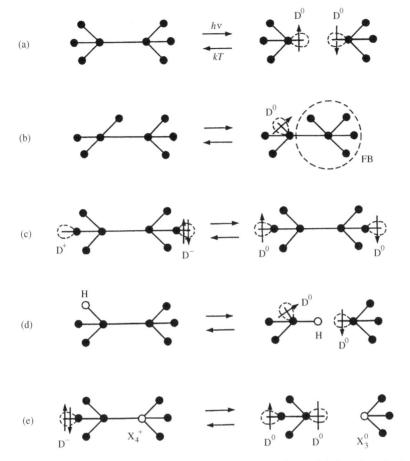

Fig. 5.6 The Staebler–Wronski effect: possible microscopic models based on intrinsic (a–d) and extrinsic (e) structural defects

negative effective electron correlation energy U_{eff} is assumed as in model (c). In addition to these models, there are further models based on charged DB pairs involving potential fluctuation within the bulk a-Si:H without assuming negative U_{eff} values [41]. In these cases, it is unrealistic to consider that the charged DB pairs are as close as those in model (c). We will now go on to discuss the various models illustrated in Fig. 5.6 in more detail.

While model (a) is a prototype for the Si–Si weak bond model, it is equipped with no mechanism to stabilize the pair of DBs (D^0s) that are produced. In this respect, model (d), which is a weak Si–Si bond model largely similar to model (a). assigns a role to the hydrogen atom whereby the two D^0s are made metastable by switching between the Si–H bond and D^0. As for the validity of model (b), which involves experimental uncertainties regarding the concept and actual existence of the floating bond (FB), no adequate evidence has yet been presented (as described in Section 3.5.2). With this in mind, only models (c), (d) and (e) will be discussed further here.

Model (d), which contains a weak Si−Si bond neighbouring a Si−H bond, has been discussed from a number of different aspects. Yonezawa and Sakamoto reported that the bond energy of the Si−Si bond is reduced by the presence of the Si−H 'back' bond, on the basis of theoretical calculations carried out using first-principle molecular dynamics [42]. It seems that this concept is valid as long as we are only considering structural models 'in the dark'. The problem with this proposal is that the two D^0s produced by the illumination are spatially located too close to each other. As mentioned above, the distance between adjacent D^0s (DBs) has been experimentally determined as $\geqslant 10$ Å, and so the main problem is how to meet this condition. Morigaki and coworkers proposed a model in which the bond switching between D^0 and Si−H is followed by repeated tunnelling of hydrogen; thus a D^0/Si−H paired unit will move increasingly further away from the other D^0 [43]. There is sill another problem to be considered, namely the spatial distance between D^0 and H. Biegelsen and Stutzmann examined the hyperfine structure of the ESR spectrum for D^0 at $g = 2.005$, using a-Si:D samples with hydrogen replaced with deuterium, and found that the hyperfine structure of a-Si:D is almost identical to that of a-Si:H [37]. This means that no hydrogen (or deuterium) exists in the immediate neighbourhood of the dangling bond (D^0). Experiments based on a pulsed ESR technique, carried out by Yamasaki et al., revealed that the distance between D^0 and H is $\geqslant 4.2$ Å [27]. This particular problem may be partially resolved by introducing a three-centerd bond (Si−H−Si) (as described in Section 3.5.1), but further experiments on this are still required [41, 42].

In model (d), the diffusion range of the hydrogen is also important. Santos et al. carried out some elaborate experiments in order to study the effects of illumination on the diffusion of hydrogen [44]. At temperatures below room temperature, it was found that the diffusion of hydrogen is extremely hindered; further studies are needed in order to provide a full analysis.

On the other hand, the effective electron correlation energy U_{eff} is assumed to be negative in models (c) and (e), as distinct from the others. Therefore, in the case of model (c), the defect reaction given by the following is exothermic [40]:

$$2D^0 \longrightarrow D^+ + D^- \tag{5.10}$$

In the above, the left-hand side represents a metastable state (as illuminated), while the right-hand side represents a stabilized state (as annealed). In this case, the effective electron correlation energy U_{eff} is given by the following:

$$U_{eff} = U_c - \frac{e^2}{\varepsilon r} - W < 0, \tag{5.11}$$

where U_c is the pure electron−electron correlation energy ($U_c > 0$), e the electronic charge, ε the dielectric constant, r the distance between D^+ and D^-, and W the lattice distortion energy needed to stabilize the charged defects (see equation (4.16) in Section 4.1.4(a)).

This model is essentially equivalent to defect model and photo-induced ESR in the chalcogenide amorphous semiconductors, as described in Section 5.2. In the case of the chalcogenide materials, the network structure is given greater flexibility, by possessing a mean coordination number $\leqslant 3$ and ionic nature to the As$-$S bonding, while the chalcogen atom has lone-pair electrons at the top of the valence band. Its optical and electrical properties support the presence of defects which have negative electron correlation energies [1, 3, 5, 6]. In contrast, Si-based materials have a more rigid structure because silicon, a group IV element, has a coordination number of 4 and its bonding lacks ionic character. Nevertheless, the presence of hydrogen will lead to a small amount of structural flexibility, as described in Section 5.2. However, in order to relate this to the fact that $U_{\text{eff}} < 0$, further studies are required. A theoretical study carried out by Bar-Yam and Joannopoulos which suggested the possibility that $U_{\text{eff}} < 0$ for the D^{0}s (DBs) in a-Si:H, is far from conclusive [45].

As mentioned above, charged DBs (D$^+$ and D$^-$) can exist under the condition where $U_{\text{eff}} > 0$, only if the potential fluctuations in the bulk are greater than U_{eff} [41]. Although the origin of the potential fluctuations is not clear, this model seems to be worthy of further consideration.

It does seem clear, however that for both of the models (c) and (d), hydrogen is be involved, either directly or indirectly, in the structural origin of the Staebler$-$Wronsky effect. In fact, a number of reports have pointed out the correlation between the Staebler$-$Wronsky effect and the content of hydrogen in a-Si:H and/or the Si$-$H bonding configuration [46, 47]. According to Nakamura *et al.*, the photodegradation is suppressed at lower temperatures as the hydrogen content is lower [46]. On the other hand, Osawa *et al.* point out that cluster-like Si$-$H or Si$-$H$_2$ bondings contribute to the photodegradation [47].

Model (e) can be regarded as an extrinsic-structural-defect model which directly involves impurity (group V) elements (represented by X in the figure) other than Si and H, although the effective electron correlation energy is negative ($U_{\text{eff}} < 0$), as in model (c). According to Mott's (8-N) rule, three fold coordination provides the highest stability in group V elements, and it also seems to give local structural freedom, together with the ionicity of the Si$-$X covalent bond. Fig. 5.6 (e) represents an example of a realistic model, but not a conclusive one. Let us consider what other elements could be impurities apart from those of group V. As the density of the DBs produced by illumination is $10^{16}-10^{17}$ cm^{-3}, the concentration of impurities must be higher than 10^{17} cm^{-3}, in order to render model (e) feasible. As described in Section 3.6, O, N and C are normally present in a-Si:H at concentrations of $\geqslant 10^{17}$ cm^{-3}. In this sense, these three elements are potential candidates for impurities in model (e)

Figure 5.7 shows the photo-created ESR spin density of a-Si:H, deposited in an ultra-high vacuum chamber, as a function of the content of impurities (N,O) [48]. As is evident from this figure, the dependence of the spin density on the N or

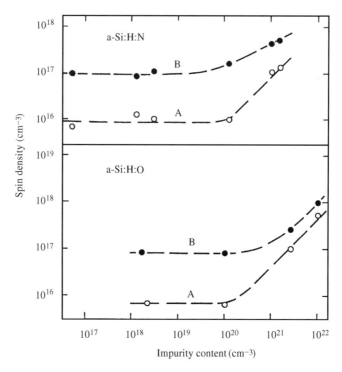

Fig. 5.7 The dependence of spin density (DB density) on the concentration of impurities (nitrogen and oxygen). (A) before illumination (in the dark); (B) after illumination [48, reproduced with permission. Copyright 1984 by the American Institute of Physics]

O content is obscured at concentrations below a certain level. In the case of nitrogen, in particular, it has been found that DBs can be induced at concentration as high as 10^{17} cm^{-3} in a-Si:H samples where the N content has been reduced to below 10^{15}/cm^{-3}, thus N seems to be excluded as a potential impurity in the extrinsic structural model [49]. For oxygen, however, there is currently no experimental evidence available which possibility decisively. Model (e) in Fig. 5.6 is essentially equivalent to the model proposed by Redfield and Bube [50].

With reference to the various microscopic structural models discussed above that have been proposed for the formation of dangling bonds under illumination, it is generally recognized among researchers that models (c), (d) and (e) seem to be the best candidates. To be more precise, none of these models can be completely rejected at the present time. The same applies to the case of the potential fluctuation model which is not shown in Fig. 5.6. It is an important feature that the photo-induced defects seems to involve not only dangling bonds, but also incorporate others defects of a totally different nature.

Figure 5.8 illustrates this point [51]. In this figure, the photoconductivity of a-Si:H (represented by the product $\mu\tau$) is plotted against the optical absorption

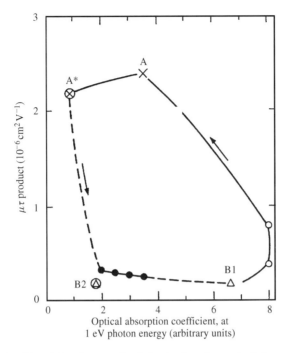

Fig. 5.8 Relationship of the photoconductivity ($\mu\tau$) to the optical absorption coefficient (α), as a function of illumination and temperature: (A* $\longrightarrow$ B2) illuminated at 160 K; (B1 $\longrightarrow$ A) annealed at 430 K [51, reproduced with permission]

coefficient α (at a photon energy of 1 eV) at both room temperature (300 K) and lower temperature (i.e. 160 K), after illumination (state B) and annealing (state A). When a sample in the annealed state (A^*) is illuminated at 160 K, $\mu\tau$ decays steeply to the point B2. Subsequently, if illumination is continued at 300 K, $\mu\tau$ becomes saturated while the absorption coefficient increases to reach point B1 (shown by the broken line). Stepwise annealing then brings the curve to point A (annealing at 430 K) and back to state A* (annealing at 480 K) (shown by the solid line). If these changes in $\mu\tau$ and α are caused by a single kind of defect, the curves should take identical courses when both ascending and descending. In fact, the courses shown by the broken and solid lines are completely different from each other. This suggests that the illumination produces at least two different types of defects [51]. It has in fact been directly demonstrated in P-doped a-Si:H (to be described later) that two or more kinds of defects are involved in these processes.

5.3.4 PHOSPHORUS-DOPED a-Si:H

As in the case of non-doped a-Si:H, photo-induced effects in a-Si:H doped with phosphorus have been studied in great detail, using deep-level transient spectroscopy (DLTS) [52], isothermal capacitance transient spectroscopy (ICTS) [53], sub-gap optical absorption measurements [28], and ESR measurements based on electron–nuclear double resonance (ENDOR) involving hydrogen (^{1}H) [27]. The most significant differences from non-doped a-Si:H include a shift of the Fermi level E_F to close to the mobility edge of the conduction band E_c by doping with P, and the admission of P into a-Si:H at controlled concentrations. E_F is located at *ca.* 0.2 to 0.4 eV below E_c, and an ESR signal of $g = 2.0055$ is not observable, which suggests that the dangling bond (D^0) normally has two electrons which are negatively charged (D^-). At the same time, as E_F is close to E_c, the dark conductivity grows and a Schottky junction is formed with the metal, thus making transient capacitance techniques such as DLTS and ICTS useful methods of analysis. In particular, the ICTS method developed by Okushi and coworkers, has proved to be a very useful techniques for exactly determining the energy distribution of the density of the localized states, thus complementing certain disadvantages of the DLTS method [54].

Figure 5.9 shows density-of-gap-states spectra ($N(E)$) of P-doped (153 ppm) a-Si:H both before and after illumination, as measured by the ICTS and photo-ICTS [54] methods [30]. A full reversibility has been confirmed. In this figure, D^- (region B) is considered to represent charged dangling bonds with two electrons, while $^*D^-$ (region A) represents charged dangling bonds coupled with four-fold coordinated phosphorus (P_4). The shallow level (region C) is identified as an electron spin center coupled with a phosphorus atom (^{31}P), which is

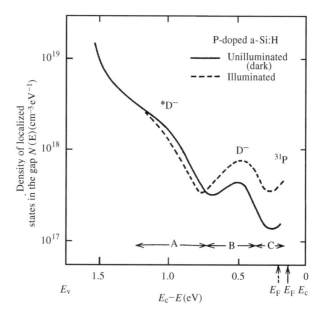

Fig. 5.9 Density of localized states in the gap of P-doped a-Si:H before and after illumination [30, reproduced with permission]

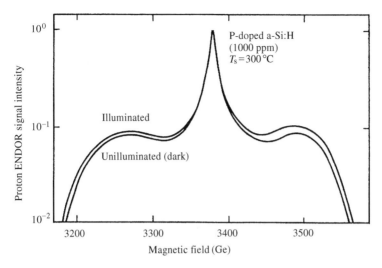

Fig. 5.10 Hyperfine splitting ($\sim$240 G) of the electron spin coupled with the phosphorus (^{31}P) nuclear spin, as measured by the ^{1}HENDOR method [27, reproduced with permission. Copyright 1987 by the American Physical Society]

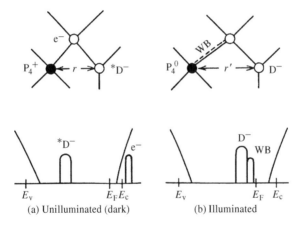

Fig. 5.11 The two-dimensional structural model used to explain the experimental data given in Figs 5.9 and 5.10 [30, reproduced with permission]

confirmed experimentally in the form of hyperfine splitting ($\sim$240 G) of the ESR center and the ^{31}P nuclear spin, as shown in Fig. 5.10 [27]. There is a possibility that this electron spin could be attributed to a donor, (see Section 3.7.3 for further details). The spin center is augmented after the illumination. The data shown in Fig. 5.9 indicate that the illumination converts a part of the $^*D^-$ into D^-, and the paramagnetic center which is coupled with the ^{31}P increases. Fig. 5.11 shows a microscopic structural model that has been used for explaining the experimental data [30]. This model takes into consideration atoms other than Si and H, and in this sense it is close to the extrinsic structural model (e) in Fig. 5.6. While this model does not require the effective electron correlation energy to be negative, the lattice relaxation W defined by equation (5.11) contributes to the reduction of energy in comparison to an ordinary D^-. The illumination induces a displacement, so that the electron takes an antibonding state of the weak Si−P bond located below E_F, thus causing hyperfine interactions with the P nuclear spin. As P then becomes neutral, the $^*D^-$ is automatically converted into an ordinary D^-. This model successfully explains the shift of E_F towards the mid-gap after illumination.

As described above, multiple structures appear in the density-of-states spectrum in the energy gap of P-doped a-Si:H and these exhibit different behaviors under illumination. In addition, paramagnetic spins are observed in direct coupling with the P atoms, and these increase or decrease depending upon the conditions of illumination and annealing. It should be noted that P-doped a-Si:H is a much more complicated material than its non-doped counterpart, thus reflecting an increased information content owing to the addition of a new 'parameter', i.e. phosphorus atoms. Although the model shown in Fig. 5.11 is far from a conclusive one, it will be considered later in conjunction with non-doped a-Si:H.

5.4 Thermal Stability

5.4.1 DIFFUSION AND DESORPTION OF HYDROGEN AND CRYSTALLIZATION OF AMORPHOUS FILMS

When a-Si:H is progressively heated, the hydrogen begins to diffuse freely within the semiconductor film. This may be likened to a hydrogen 'liquid' within the rigid framework of the Si random network, and Street *et al.* have named this system as a 'hydrogen glass', with the temperature at which the hydrogen begins to diffuse freely being known as the equilibrium temperature T_E of the glass [55]. This can be regarded as a form of the defect formation model described in Section 3.5.4, where the thermal equilibrium process is directly linked with the hydrogen, and T_E may be thought of as the glass transition temperature of the hydrogen glass [55]. Matsuo *et al.* have confirmed this behavior through the use of thermal analysis [56]. The diffusion of hydrogen can be facilitated as the density of the dangling bonds is greater, and the levels of impurities such as P and B are higher. In this model [55], it is assumed that hydrogen diffuses by a hopping process, always taking the route along the Si−H or P−H bonds, while constantly changing the coordination number of the impurities and the number of D^0s [55].

Figure 5.12 shows the dark conductivity−temperature characteristics of P-doped a-Si:H [55]. The samples are quenched down to room temperature from 200 °C at various cooling rates, and the dark conductivity is then measured as the samples are heated. It should be noted that the data points for all of the samples lie on a single line at temperatures higher than $T_E = 130$ °C, a behavior which

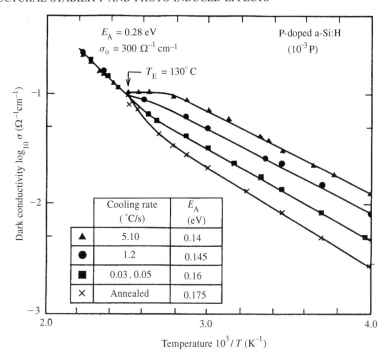

Fig. 5.12 Dark conductivity–temperature characteristics of P-doped a-Si:H, as a function of the cooling rate [55, reproduced with permission. Copyright 1987 by the American Physical Society]

closely resembles that of the normal glass transition phenomenon. The 'glass transition temperature' (T_E) is *ca.* 80°C for B-doped samples, and higher than 150 °C for non-doped samples.

Figure 5.13 shows the temperature dependence of the diffusion coefficient D of hydrogen as determined by using deuterium [55]. In this figure, the data for non-doped a-Si:H have been taken from the work of Carlson and Magee [57]. The diffusion coefficients in P- and B-doped samples are substantially greater than that in non-doped samples. In particular, values measured for B-doped a-Si:H are very high. In the case of compensated a-Si:H doped with both P and B, the diffusion coefficient is reduced to a value which is as low as that found in non-doped samples.

In the case of ordinary glass materials, the transition to the glass state occurs normally at a viscosity of around 10^{13} P (dyn. s/cm^2). For covalently bonded glasses such as SiO$_2$, the viscous flow can be directly linked with bond rearrangements which are implemented through a mechanism related to the self-diffusion of the elements which constitute the glass. The viscosity η is related to the diffusion coefficient D through the Stokes–Einstein relation, as follows:

$$\eta D = kT/6\pi R, \tag{5.12}$$

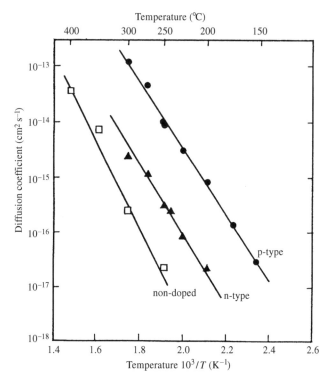

Fig. 5.13 Diffusion coefficient of deuterium in amorphous silicon as a function of temperature [55, reproduced with permission. Copyright 1987 by the American Physical Society]

where R is the effective radius of the self-diffusing species. If T is equal to the equilibrium temperature T_E ($\sim 100\,°C$), and $R = 3$ Å, the right-hand side of equation (5.12) has a value of *ca.* 10^{-7} dyn, and the diffusion coefficient at the glass transition temperature is estimated to be *ca.* 10^{-20} cm² s⁻¹. On the other hand, the equilibrium temperatures determined by extrapolating the curves for the measured data shown in Fig. 5.13 to 10^{-20} cm² s⁻¹ is 130 °C for n-type (P-doped) samples, and 80 °C for p-type (B-doped) materials where the former value is in good agreement with the T_E shown in Fig. 5.12. This clearly demonstrates that the equilibrium phenomenon seen in the temperature cycle of the dark conductivity shown in Fig. 5.12 is attributable either to changes in the viscosity of the hydrogen glass or to the diffusion of hydrogen.

If the temperature is raised further, hydrogen starts to be desorbed out of the a-Si:H film. The process of hydrogen desorption has been previously discussed in detail in Section 3.2.3.

When the temperature exceeds 600 °C, amorphous films begin to crystallize into c-Si, leaving just a minute amount of hydrogen in the film. In contrast to glass materials prepared by liquid phase quenching, the concepts of a supercooled liquid phase and a glass transition are not clearly definable in the random network of Si. Moreover, the crystallization temperature is considerably affected by the nature of the substrate materials, the degree of film–substrate adherence, the substrate temperature at the time of preparation, the film thickness, and various other factors. In terms of thermodynamics, the crystallization process is significantly affected not only by the bulk (volume) free energy, but also by the surface and/or interfacial free energies.

According to Tsai and Fritzsche, when non-doped a-Si:H films are heated at a rate of $25\,°C\,min^{-1}$ crystallization occurs at temperatures between 680 and $780\,°C$, with the release of 1.7–$2.2\,kcal\,mol^{-1}$ of heat [58].

The crystallization temperature T_c of P-doped a-Si:H is determined by measuring the activation energy of the dark conductivity [59]. In this case, as the heating rate is as low as $1.5\,°C\,min^{-1}$, the crystallization temperature T_c is expected to be somewhat lower than that determined in the previous experiments. Figure. 5.14 shows the dependence of the crystallization temperature T_c of a-Si:H on the level of P-doping, as determined in this way [59]. As the content of P is increased, the crystallization temperature rises slightly at first but then drops steeply to $500\,°C$. As far as this present author is aware, no further detailed data are available on this crystallization process.

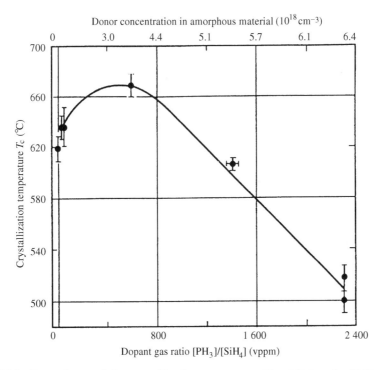

Fig. 5.14 Dependence of the crystallization temperature T_c of P-doped a-Si:H on the doping level [59, reproduced with permission]

5.4.2 CRYSTALLIZATION PROCESSES IN MULTILAYER FILMS

Multilayer films and superlattices are topics which are currently attracting great interest in the field of amorphous semiconductors [60–62]. While there are a few reports on the thermal stability of amorphous multilayer films, data obtained for the crystallization temperature vary widely depending upon the system conditions [63–67]. For example, the crystallization temperature T_c of an a-Ge film sandwiched between lead (Pb) films is considerably lower than that of single a-Ge film [63], while in the case of an a-Ge film sandwiched between a a-GeN films, T_c is much higher than that of a single a-Ge film [64]. This may be attributable to a dramatic change in the nature of the interfacial energy, which depends upon the particular combination of the film materials. A qualitative explanation is attached below by using a simple thermodynamic model.

Let us assume that the crystalline nuclei, the so-called embryos, are formed at a certain temperature in the supercooled liquid phase. According to the well-known theory of homogeneous nucleation, an embryo can survive as a stable nucleus of crystallization only if its size exceeds a certain critical level, thus spreading the crystallization to the whole system so as to minimize its free energy [68]. If a nucleus is of a spherical form, the radius of its critical size r_c is given by the following [68, 69]:

$$r_c = 2\gamma/\Delta G_V \qquad (5.13)$$

where γ represents the free energy per unit area of the liquid/crystal interface, and ΔG_V is the decrease in the bulk free energy per unit volume caused by crystallization of the supercooled liquid. A similar argument applies to film growth on a substrate, which is referred to as inhomogeneous nucleation by cap-shaped clusters [69].

As an application of this theory, let us consider the crystallization process of an amorphous semiconductor film A sandwiched between the materials B, as shown in Fig. 5.15(a), where material B is assumed to have sufficient thermal stability and the interlayer atomic diffusion of the constituent elements across the AB interface can be ignored. While the crystallization temperature T_c in the homogeneous nucleation model can be determined by a simple method as a function of the critical radius (r_c) of the spherical embryos generated in the film, it may be affected in a different way if the film thickness d is smaller than r_c. As illustrated in Fig. 5.15, the crystalline nucleus can be regarded as a cylinder of radius r to a first-order approximation.

The change in the free energy ΔG of the system where a cylindrical crystalline nucleus is formed in a two-dimensional film is given by the following expressions:

$$\Delta G = -\pi r^2 d \Delta G_V + 2\pi r^2 \Delta\gamma + 2\pi r d\gamma_{ac} \qquad (5.14)$$

with

$$\Delta\gamma = \gamma_{bc} - \gamma_{ba} \qquad (5.15)$$

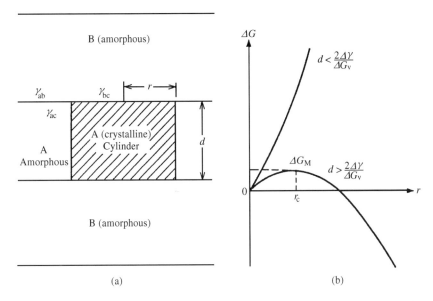

Fig. 5.15 A crystallization model for the multilayer film (with layer A sandwiched between the layers of B): (a) illustration of the cylindrical crystalline nucleus; (b) Change in the free energy as a function of cylinder radius [66, reproduced with permission]

where ΔG_V is the increment of the free energy per unit volume caused by the formation of the cylindrical crystalline nucleus, $\Delta \gamma$ the increment of the interfacial free energy per unit area of the AB interface caused by crystallization of the film A, γ_{ac} the interfacial free energy per unit area of the crystalline/amorphous interface in material A (film), and γ_{ba} represents the interfacial free energy per unit area between the material B and the initial amorphous material A and γ_{bc} that between the material B and the crystallized material A [66, 67].

As is evident from equation (5.14), if $d < 2\Delta\gamma/\Delta G_V$, then ΔG continues to increase with r, and if $d > 2\Delta\gamma/\Delta G_V$, ΔG takes a maximum value ΔG_M at $r = r_c$. (see Fig. 5.15(b)). Here, r_c and ΔG_M are given by the following expressions:

$$r_c = d\gamma_{ac}/(d\Delta G_V - 2\Delta\gamma) \tag{5.16}$$

$$\Delta G_M = \pi(d\gamma_{ac})^2/(d\Delta G_V - 2\Delta\gamma) \tag{5.17}$$

Both r_c and ΔG_M change as functions of the thickness d of the film A, as shown in Fig. 5.16. If $\Delta\gamma > 0$, both r_c and ΔG_M increase as d is reduced, becoming infinitely large at $d = 2\Delta\gamma/\Delta G_V$. If $\Delta\gamma < 0$, both r_c and ΔG_M decrease as d is reduced, i.e. not only the film thickness d, but also $\Delta\gamma$ significantly affect r_c and ΔG_M. If it is assumed that the crystallization process of thin film A is mainly determined by the nucleation rate, then the amorphous–crystalline transition temperature T_c can be roughly evaluated from ΔG_M, by defining the

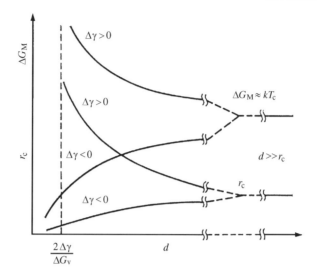

Fig. 5.16 Dependence on the film thickness d of the maximum value of the free energy change ΔG_M and the critical radius of the cylindrical crystalline nucleus r_c, with $\Delta\gamma$ as a parameter representing the interfacial free energy properties [66, reproduced with permission]

thermal activation energy for the nucleation rate by the following expression:

$$kT_c \propto \Delta G_M \qquad (5.18)$$

where k is the Boltzmann constant.

The dependence of T_c on the film thickness d may therefore be qualitatively approximated by that of ΔG_M. In the case where $\Delta\gamma > 0$, the crystallization temperature T_c rises as d decreases, while if $\Delta\gamma < 0$, then T_c drops as d decreases. When the film thickness d is reduced, the properties of the hetero-interface ($\Delta\gamma$) in the multilayer film affect the thermal stability (crystallization temperature T_c) substantially.

As is evident from this argument, the thermal stability of material A sandwiched between the materials B in a multilayer film is considerably affected by the properties of the A/B hetero-interface, if the film thickness is smaller than the radius of the critical radius of the crystalline nucleus in the bulk. If $\gamma_{bc} > \gamma_{ba}$ (i.e. $\Delta\gamma > 0$), its thermal stability is enhanced, and if $\gamma_{bc} < \gamma_{ba}$ (i.e. $\Delta\gamma < 0$), then the thermal stability is reduced.

The above discussion is based on likening the amorphous semiconductor film at high temperatures to the supercooled liquid phase. Although this is a somewhat crude analogy, it may be suitable enough for understanding the thermal stability of the multilayer film.

Figure 5.17 shows the dependence of the crystallization temperature T_c of a-Ge in a-Ge/a-GeN multilayer films as a function of the a-Ge film thickness [66, 67]. In this figure, data for a-Ge/Pb multilayer films are also included [63].

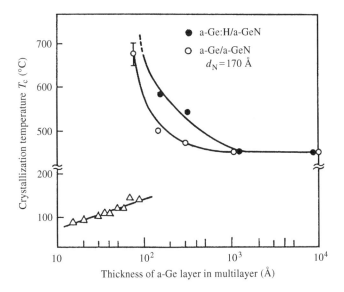

Fig. 5.17 Dependence on the film thickness of the crystallization temperature T_c of a-Ge in different multilayer films: (●) a-Ge:H/a-GeN; (○) a-Ge/a-GeN; (△) a-Ge/Pb; [67, reproduced with permission. Copyright 1989 by the American Institute of Physics]

T_c is considerably affected by the nature of the material B, i.e. a-GeN or Pb. In the case of a-GeN ($\Delta\gamma > 0$), T_c rises as the film thickness is reduced, and crystallization does not occur at thicknesses less than 100 Å. On the contrary, if a-Ge is sandwiched between Pb films, T_c is markedly lowered as the film thickness is reduced ($\Delta\gamma < 0$).

According to Commarata and Greer, if the combination of materials A and B represents a completely soluble system, the interdiffusion is accelerated across the interface and crystallization occurs readily ($\Delta\gamma < 0$) [70]. This also applies to combinations of Ge/Pb and Ge/Si. On the other hand, in the cases of Ge/GeN and Si/SiN, which do not form solid solutions, the thermal stability of a-Ge and a-Si is high enough to suppress crystallization ($\Delta\gamma > 0$). Naturally, the stability of the multilayer films is also enhanced.

In order to advance this argument further, more research is required on the structural characterization of the interface at the atomic level, along with the corresponding of theoretical studies.

References

1. K. Tanaka, *Fundamental Physics of Amorphous Semiconductors*, edited by F. Yonezawa, Springer-Verlag, Berlin, (1981) 104.
2. K. Tanaka, *Proceedings of the 14th International Conference on Amorphous Semiconductors (ICAS-14)*, edited by G. Bauer, W. Fuhs and L. Ley, North-Holland, Amsterdam (1991) 1.

3. S. G. Bishop, U. Strom and P. C. Taylor, *Phys. Rev. Lett.*, **34** (1975) 1346.
4. D. K. Biegelsen and R. A. Street, *Phys. Rev. Lett.*, **44** (1980) 803.
5. R. A. Street and N. F. Mott, *Phys. Rev. Lett.*, **35** (1975) 1293.
6. M. Kastner, D. Adler and H. Fritzsche, *Phys. Rev. Lett.*, **37** (1976) 1504.
7. P. W. Anderson, *Phys. Rev. Lett.*, **34** (1975) 953.
8. J. Hautala, W. D. Ohlsen and P. C. Taylor, *Phys. Rev.*, **B38** (1990) 9766.
9. P. J. Gaczi and H. Fritzsche, *Solid State Commun.*, **38** (1981) 23.
10. P. C. Taylor, U. Strom and S. G. Bishop, *Solar Energy Mater.*, **8** (1982) 23.
11. D. L. Staebler and C. R. Wronski, *Appl. Phys. Lett.*, **31** (1977) 292.
12. H. Dersch, J. Stuke and J. Beicher, *Phys. Status. Solidi. (b)*, **105** (1981) 265.
13. S. Yamasaki, H. Okushi, A. Matsuda, K. Tanaka and J. Isoya, *Phys. Rev. Lett.*, **65** (1990) 756.
14. S. G. Bishop, U. Strom and P. C. Taylor, *Phys. Rev.*, **B15** (1977) 2278.
15. J. Spaeth, *Semi-insulating III–V Materials*, edited by H. Kukimoto and S. Miyazawa, Ohmsa and North-Holland, Amsterdam (1986) 299.
16. F. Seitz, *Rev. Mod. Phys.*, **26** (1954) 7.
17. J. C. Phillips, *Bonds and Bands in Semiconductors*, Academic Press, New York, (1973).
18. B. Holm, K. B. Nielsen and B. B. Nielsen, *Phys. Rev. Lett.*, **66** (1991) 2360.
19. T. Sakaki and H. Katayama-Yoshida, *J. Phys. Soc. Jpn*, **58** (1989) 1685.
20. Y. Toyozawa, *Proceedings of the US–Japan Seminar on Atomic Processes Induced by Electronic Excitations in Non-Metallic Solids*, edited by W. B. Fowler and N. Itoh, World Scientific, Singapore, (1990) 3.
21. Y. Shinozuka and T. Toyosawa, *Kotai Buturi*, **14** (1979) 526 (in Japanese).
22. M. Stutzmann, D. K. Biefelsen and R. A. Street, *Phys. Rev.*, **B35** (1987-I) 5666.
23. Y. Shinozuka, (personal communication).
24. W. Fuhs, M. Milleville and J. Stuke, *Phys. Status. Solidi. (b)*, **89** (1978) 495.
25. J. J. Pankove and J. E. Berkeyheiser, *Appl. Phys. Lett.*, **37** (1980) 7056.
26. M. Stutzmann, *Phys. Rev.*, **B34** (1986) 63.
27. S. Yamasaki, S. Kuroda and K. Tanaka, *AIP Conf. Proc.*, **157** (1987) 9.
28. K. Tanaka, H. Okushi and S. Yamasaki, *Tetrahedrally-Bonded Amorphous Semiconductors* edited by D. Adler and H. Fritzsche, Plenum Press, New York (1985) 239.
29. H. Fritzsche, *Solar Energy Mater.* **3** (1980) 447.
30. H. Okushi, R. Banerjee and K. Tanaka, *Amorphous Silicon and Related Materials*, edited by H. Fritzsche, World Scientific Singapore, (1988) 657.
31. S. Nakamura, K. Watanabe, M. Nishikuni, Y. Hishikawa, S. Tsuda, H. Nishiwaki, M. Ohnishi and Y. Kuwano, *J. Non-Cryst. Solids*, **59/60** (1983) 1139.
32. M. Stutzmann, W. B. Jackson and C. C. Tsai, *Phys. Rev.*, **B32** (1985) 23.
33. M. Hack, S. Guha and W. den Boer, *Phys. Rev.*, **B15** (1986) 2512.
34. W. B. Jackson and J. Kakalios, *Phys. Rev.*, **B37** (1988) 1020.
35. D. Redfield and R. H. Bube, *Appl. Phys. Lett.*, **54** (1989) 1037.
36. H. R. Park, J. Z. Liu and S. Wagner, *Appl. Phys. Lett.*, **55** (1989) 2658.
37. D. K. Biegelsen and M. Stutzmann, *J. Non-Cryst. Solids*, **77/78** (1985) 703.
38. M. Stutzmann, *Oyo-Buturi*, **60** (1991) 1004 (in Japanese).
39. S. T. Pantelides, *Phys. Rev.*, **B36** (1987) 3479.
40. D. Adler, *J. Phys. (Paris)*, **42** (1981) C4-3.
41. H. M. Branz, *Phys. Rev.*, **B41** (1990) 7887.
42. F. Yonezawa and S. Sakamoto, *Oyo-Buturi*, 60 (1991) 984 (in Japanese).
43. K. Morigaki, *Jpn. J. Appl. Phys.*, **27** (1988) 163.
44. P. V. Santos, N. M. Johnson and R. A. Street, *Phys. Rev. Lett.*, **67** (1991) 2686.
45. Y. Bar-Yam and D. Joannopoulos, *J. Non-Cryst. Solids*, **77/78** (1985) 99.

46. N. Nakamura, T. Takahama, M. Isomura, M. Nishikuni, K. Yoshida, S. Tsuda, S. Nakano, M. Ohnishi and Y. Kuwano, *Jpn. J. Appl. Phys.*, **28** (1989) 1762.
47. M. Ohsawa, T. Hama, T. Akasaka, T. Ichimura, H. Sakai, S. Ishida and Y. Uchida, *Jpn. J. Appl. Phys.*, **24** (1985) L 838.
48. C. C. Tsai, M. Stutzmann and W. B. Taylor, *AIP Conf. Proc.*, **120** (1984) 242.
49. T. Kurihara, *Proceedings of the 50th Meeting of the Japanese Society of Applied Physics*, No. 2 (1989), 709 (in Japanese).
50. D. Redfield and R. H. Bube, *Phys. Rev. Lett.*, **65** (1990) 464.
51. D. Han and H. Fritzsche, *J. Non-Cryst. Solids*, **59/60** (1983) 397.
52. D. V. Lang, J. D. Cohen and J. P. Harbison, *Phys. Rev.*, **B25** (1982) 5285.
53. H. Okushi, M. Itoh, T. Okuno, Y. Hosokawa, S. Yamasaki and K. Tanaka, *AIP Conf. Proc.*, **120** (1984) 250.
54. H. Okushi, *Phil. Mag.*, **B52** (1985) 33.
55. R. A. Street, J. Kakalios, C. C. Tsai and T. M. Hayes, *Phys. Rev.*, **B35** (1987-II) 1316.
56. S. Matsuo, H. Nasu, C. Akamatsu, T. Imura and Y. Osaka, *MRS Symp. Proc.*, **118** (1988) 297.
57. D. E. Carlson and C. W. Magee, *Appl. Phys. Lett.*, **33** (1978) 81.
58. C. C. Tsai and H. Fritzsche, *Solar Energy Mater.*, **1** (1979) 29.
59. O. J. Reilly and W. E. Spear, *Phil. Mag.*, **B38** (1978) 295.
60. B. Abeles and T. Tiedje, *Phys. Rev. Lett.*, **51** (1983) 2003.
61. S. Miyazaki, Y. Ihara and M. Hirose, *Phys. Rev. Lett.*, **59** (1987) 125.
62. K. Hattori, T. Mori, H. Okamoto and Y. Hamakawa, *Phys. Rev. Lett.*, **60** (1988) 825.
63. L. A. Clevenger, C. V. Thompson and R. S. Cammarata, *Appl. Phys. Lett.*, **52** (1988) 795.
64. I. Honma, H. Hotta, K. Kawai, H. Komiyama and K. Tanaka, *J. Non-Cryst. Solids*, **97/98** (1987) 947.
65. L. Ley, *MRS Symp. Proc.*, **118** (1988) 329.
66. K. Tanaka, I. Homma, H. Tamaoki and H. Komiyama, *MRS Symp. Proc.*, **118** (1988) 329.
67. I. Honma, H. Komiyama and K. Tanaka, *J. Appl. Phys.*, **66** (1989) 1170.
68. M. Volmer and A. Wever, *Z. Phys. Chem.*, **119** (1925) 227.
69. A. Kinbara and H. Fujiwara, *Thin Films*, Shokabo, Tokyo (1979) 53 (in Japanese).
70. R. C. Commarata and A. L. Greer, *J. Non-Cryst. Solids*, **61/62** (1984) 889.

6 APPLICATIONS

In this chapter, we will describe the various features of amorphous semi-conductors from the viewpoint of their applications, in particular those newer fields of application which take advantage of the specific merits of these materials, and present overviews of a number of typical examples. Detailed descriptions of the principles and characteristics of some typical examples of practical application will be given, including solar cells, color sensors, one-dimensional optical sensors for facsimile devices and X-ray computerized tomography, two-dimensional image sensors, field-effect transistors for liquid crystal displays, photoreceptor drums for copying machines and laser beam printers, and ultrasensitive image pick-up tubes based on amorphous selenium (a-Se).

6.1 Features and Applications of Amorphous Semiconductors

Among the variety of amorphous semiconductors that have been investigated, those based on the chalcogenide and tetrahedral materials are the ones mainly used for applications in functional devices. Some examples of these applications are summarized in Table 6.1. In this table, examples (1)–(5) refer to selections for non-volatile memory devices (where the stored data remains intact when the power is turned off), based on atomic displacement effects. Chalcogenide materials are mainly used for this purpose. Examples (6)–(14) are characterized by the common use of electronic functions. Example (14) represents an attempt to synthesize new materials under artificial structural control and to derive innovative functions which do not exist in nature. For the latter purpose, materials based on tetrahedral structures are often used, while for electronic copying machines and image pick-up tubes, chalcogenide materials are also utilized. Let us now see which particular advantages of the amorphous semiconductors are utilized in these applications.

(1) *Large-area, thin-film and multilayer preparations are readily available.* Amorphous semiconductors can be prepared by various methods, such as quenching from the liquid state (one of the earliest techniques), transforming crystalline materials into amorphous ones by irradiating with high-energy particles (e.g. the implantation of high-energy ions), sputtering of solid materials, and thermal, radio-frequency or photochemical vapour deposition (CVD) from the gas phase.

Table 6.1 A summary of the functions and device applications of amorphous semiconductors

Features	Target function	Material		Utilized function	Applications	Category[a]
		Chalcogenide	Tetrahedral			
Structural changes	(1) Thermal shape change	Se; As−Te−Se	—	Shape change such as perforation by laser irradiation	Irreversible/reversible optical memory	NVM
	(2) Amorphous−crystalline phase change	Se−Te; Ge−As−Te; Sn−Te−Se; Te−As−Si−Ge	—	Phase change induced by electrons, light or electrical current	High-density rewritable optical memory; laser-beam printer; electron-beam memory	NVM
	(3) Photodoping	Ag/As−S−Te; Ag/Se−Ge	—	Optical parameters changed by shift of metal atoms into amorphous layers through light pulses.	Image memory; resists; printing	NVM
Structurally induced electronic state changes	(4) Photostructural change	As−Se; As−S; Ge−S; Ge−Se; Se−As−Ge; Ge−As−S−Se	a-Si:H	Changes in optical parameters, conductivity, and chemical stability by illumination	Optical switch; optical memory	NVM
	(5) Photostopping	As−Se	—	Changes in photo-transmittance with defects controlled by illumination with shorter wavelengths	Optical switch; optical memory	NVM

(continued overleaf)

Table 6.1 (*continued*)

Features	Target function	Material		Utilized function	Applications	Category[a]
		Chalcogenide	Tetrahedral			
Changes in electronic state and transport phenomena	(6) Distortion effect	—	a-Si:H; μc-Si:H	Changes in conductivity by distortion	Distortion sensor	BEE
	(7) Thermoelectric effect	—	a-Si:H	Temperature-difference electromotive force in bulk or junction	Radio-frequency power sensor	BEE
	(8) Magnetic effect	—	—	Hall effect; magnetoresistance effect	Hall device; magnetoresistance device	BEE
	(9) Photoconductive effect	Se, Se–Te, Se–As, Se–Te–As	a-Si:H (addition of C, N, O, Ge)	Changes in conductivity caused by carrier production or mobility changes by optical absorption	Electrophotograph; laser-beam printer; image tube; image device; photosensor; radiation sensor	PEE
	(10) Photovoltaic effect	—	a-Si:H (addition of C, N, O, Ge), μc-Si:H	Separation of photocarrier by p/n, p/i/n, Schottky or MIS junction	Solar cell; photosensor	PEE

(11)	Electric-field effect	—	a-Si:H	Conductivity control by field, accumulation and transfer of carriers	Thin film transistor; charge-coupled device; 2D-, 3D-IC	EF/IE
(12)	Non-linear effect	Te–Ge–Sb–S, Te–As–Si, Se	a-Si:H	Non-linear response to electric field and light	Avalanche multiplication image tube; double-implant device; varistor	EF/IE
(13)	Ambipolar carrier implanting	—	a-Si:H, a-SiC:H	Control of minority carrier injection; emission by electron–hole injection; recombination rate control	Bipolar transistor; light-emitting diode; 2D-, 3D-IC; heteroemitter	EF/IE
Artificially induced structural effect	(14) Ultra-thin, low-dimensional effect	—	a-Si:H (addition of C, N, O, Ge); μc-Si:H; porous Si; ultrafine particles	Quantum effects, such as tunnel and miniband; stacking effects, such as carrier separation and superdoping; low-D effects such as enhanced mobility; temperature-characteristic control	Solar cell; thin film transistor; phosphor; superstructure devices, such as light-emitting diode	J/IQE

[a]NVM, non-volatile (stable) memory; BEE, bulk electron effect; PEE, photo-electric effect; EF/IE, electric-field, injection effect; J/IQE, junction, interface quantum effect

Since most amorphous semiconductor devices are fabricated from thin films, CVD is most frequently used for their preparation. To be more specific, amorphous semiconductor materials based on chalcogen elements such as S, Se, As and Te are transformed into devices via the process of film deposition, beginning with the vaporization of the solid materials by vacuum evaporation and sputtering.

On the other hand, tetrahedral materials, typically represented by amorphous silicon, can be readily processed into large-area thin film devices through the use of glow discharge, and photo- or thermal decomposition of source gases such as monosilanes. For example, the continuous production of a-Si:H solar cells, tens of centimeters in width and hundreds of metres in length, has been achieved through the use of the glow discharge system illustrated in Fig. 6.1. In this arrangement, a continuous seamless multilayer films consisting of SiO_2, stainless steel, p-type a-Si:H, i-type a-Si:H, and n-type a-Si:H, with a transparent electrode on the stainless steel substrate, can be produced in specified patterns, ready for cutting and stacking. Such a continuous mass production process would be impossible for crystalline materials. Moreover, as devices of desired geometry can be fabricated by patterning, and stacking with heterogeneous materials can be implemented, designs with complicated functions are possible.

(2) *Low-temperature fabrication using a variety of substrate materials, is possible.* The substrate temperature during film preparation may be as high as 500°C, but is normally ⩽300°C thus allowing the use of conventional glasses, metal films (such as aluminum and stainless steel), or polymer films. This allows the manufacture of flexible devices at a lower cost and with a high degree of geometrical variation, as shown in Fig. 6.1. Owing to fabrication lower temperatures, interlayer reactions in multilayer preparations and the diffusion of impurities are limited, and precise control at the monoatomic-layer level is possible for constructing multilayer structures. Furthermore, as the energy demands for substrate heating are reduced, less energy is required in fabrication (environmentally friendly process).

(3) *A wide range of control of the various properties is possible by alloying and doping with impurities.* The optical gap and conductivity of chalcogenide-based amorphous materials can be controlled by multicomponent-alloying with S, Se, As, Te, Si, Ge and Sb, or by doping with transition elements. In the case of tetrahedral materials, this can be achieved by alloying silicon with C, Ge, Sn, N and O, or by doping with valence-control impurities. The composition can be continuously adjusted, thus presenting a greater degree of freedom in structural control. In particular, the advent of hydrogenated amorphous silicon (a-Si:H) has made it possible to realize p- and n-control by doping with B or P, as in crystalline semiconductors, which greatly expands the range of possible applications. It seems to provide larger non-linearity as compared to crystalline.

In contrast, the following may be regarded as disadvantageous properties of those materials.

(4) *Lack of long-term stability.* Since the amorphous structure is in a non-equilibrium state, it lacks long-term stability. It can be crystallized over an infinitely

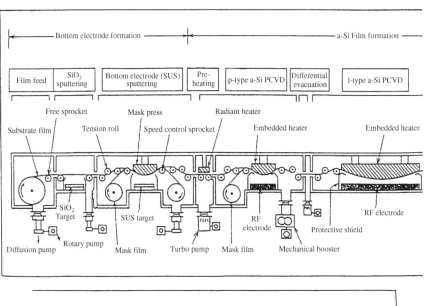

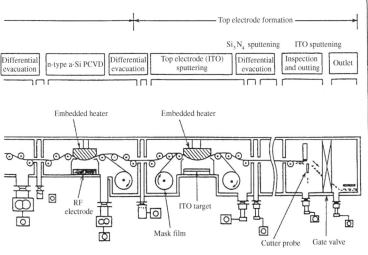

Fig. 6.1 Schematic representation of a continuous film producing system used for fabricating large-area solar cells (reproduced by permission of Hitachi Ltd)

long time-scale. However, this may provide no actual problems in practice, as witnessed by Mesopotamian glassware, the majority of which has kept its amorphous structure for over 6000 years. Hydrogenated amorphous silicon is subject to certain photostructural changes known as the Staebler–Wronski (SW) effect, as previously described in Chapter 5, which can cause serious concern in the case of those applications where devices are exposed to intensive exposure to light for long periods of time, e.g. as solar cells for power generation. While the detailed mechanism of these structural changes is not yet fully understood, some causes have been identified, and various means for stabilizing the materials have been systematically investigated. In some cases, stable utilization has been achieved in practice through the ingenious design of device structures.

(5) *Low mobility of carriers.* In amorphous semiconductors, the mobility of the carriers, i.e. electrons or holes, is lower than that of the crystalline semiconductors by approximately two orders of magnitude (or even more), which is disadvantageous in their applications in high-speed or high-frequency devices. However, this low mobility may be regarded as being equivalent to high impedance, and thus can be effectively utilized in reduced-power-consuming devices. This feature has already been applied commercially in thin film transistors for powering liquid crystal flat-panel displays and photosensitive drums for electron photography (to be described below). In solar cell applications of amorphous Si, the (disadvantage of a lower mobility than that of the crystalline material is compensated by the advantage of a photoabsorption coefficient which is larger than that of the latter material by an order of magnitude or more; this also leads to savings in materials requirements.

As mentioned in above, amorphous materials have a number of attractive properties distinct from those of crystalline materials, and a variety of devices have been proposed by utilizing these. Some examples of the major applications will be described below.

6.2 Examples of Applications of Amorphous Semiconductors

6.2.1 SOLAR CELLS

Following the report of Spear and coworkers on valence control in amorphous Si (see Chapter 1), Carlson and Wronski fabricated Pt/a-Si:H Schottky-type junctions and pointed out the feasibility of their application to solar cells or photovoltaic (PV) cells [1]. Hamakawa and coworkers reported the preliminary manufacture of (p-i-n)-type a-Si cells which currently constitute a standard solar-cell design [2]. These solar cells were first commercialized as a power source for table-top calculators by the Sanyo Electric Company and the Fuji Electric Company [3]. This was the first commercial product containing amorphous Si, displaying a low power consumption for the calculator's liquid crystal display, operation under artificial illumination such as fluorescent lighting, and good matching with the spectral response of a-Si solar cells.

Since the optical absorption coefficient of amorphous Si is much larger than that of crystalline Si, film materials as thin as 1 μm are suitable; these can be prepared at lower temperatures. From the viewpoint of resources, Si is one of most abundant elements on earth. These facts identify amorphous silicon as the most promising candidate for use in solar cells in the future from the point of

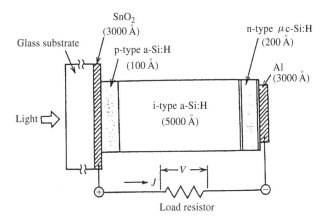

Fig. 6.2 Schematic representation of a cross-section through an amorphous Si solar cell

view of power generation. In the face of the energy crisis triggered off by the so-called petroleum crisis, and continuing global environmental problems, research efforts have been concentrated on the implementation of solar cells based on amorphous Si as a national project in Japan.

Figure 6.2 shows a cross-sectional structure of a (p-i-n)-type amorphous Si solar cell. Other types of trial-manufactured PV cells, such as Schottky and MIS cells, are much inferior to the (p-i-n)-type cells in their photovoltaic conversion efficiencies. The (p-n)-type cells based on amorphous Si have poorer junction properties than those based on crystalline Si, and hence the conversion efficiency of the former is lower than its crystalline counterpart. This may be attributed to the reduction of the effective optical gap caused by doping with B (as a p-type impurity) or P (as an n-type one), which not only leads to active coordination as acceptors or donors but also creates recombination centers at deeper levels within the gap. For this reason, the recombination rate of the p-n junction becomes fast enough to augment the conductivity in the impurity levels, thus resulting in essentially Ohmic-junction properties. The construction of a (p-i-n) structure results in a situation where the i-layer absorbs light energy to create electron–hole pairs, and the p- and n-layers set Fermi levels in the neighbourhood of the valence and conduction bands, respectively, thus creating internal electric fields for carrying the electrons and holes from the i-layer to either electrode. In this way, the cell properties are improved by the division of roles, thus taking advantage of the merits of each layer.

Hamakawa and Okamoto successfully improved the conversion efficiency by reducing the recombination that occurs at the surface layer through the use of B-doped a-SiC:H, for which the optical gap is wider than that of B-doped a-Si:H in the p-layer on the incident-light side [4]. Further improvements to the cell properties, particularly the conversion efficiencies, have been attempted through the use of P-doped microcrystalline (μc) Si:H of higher conductivity in the n-layer and rough-faced transparent electrodes in order to enhance the optical absorption coefficient by confining the incident light.

Figure 6.3 shows the dark current–voltage characteristics of a (p-i-n)-type diode which presents good rectifying properties. In addition, Fig. 6.4 shows the current–voltage characteristics of an amorphous silicon solar cell under AM–1.5

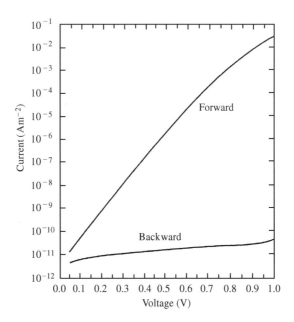

Fig. 6.3 Rectifying characteristics of the (p-i-n)-type diode

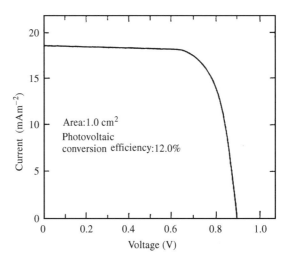

Fig. 6.4 Output characteristics of the amorphous Si solar cell

light irradiation conditions, for which the photovoltaic conversion efficiency is better than 10%.

The intensity of solar irradiation in the bright sky directly on the equator is about 100 mW/cm^{-2}, and its energy is distributed over a broad spectral range (as shown in Fig. 6.5). The spectral sensitivity of the amorphous Si solar cell a quantum efficiency of *ca* displays 90% at around 500 nm, as shown in this figure, while only the light energy over the very limited range of 400 to 700 nm is utilized. It should be noted that the utilization of energy at longer wavelength, in particular, is very low.

When utilizing the photo-electromotive effect at the (p-i-n) junction for the photovoltaic conversion of light having a broader spectral distribution, such as solar radiation, the reduction of the optical gap allows the absorption of light, including the longer wavelength range, to enhance the output current. However, this may reduce the potential difference at the junction which depends upon the optical gap, and thus could lower the output voltage. If the optical gap is increased, the output current then becomes smaller and the output voltage is increased.

With the spectrum of solar radiation (as shown in Fig. 6.5), the output power (= current × voltage) reaches a maximum at an optical gap of *ca* 1.5 eV. The

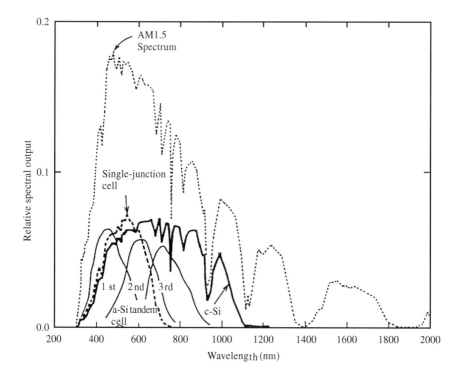

Fig. 6.5 Solar radiation spectrum and the spectral responses of various types of solar cell

optical gap of amorphous Si is around 1.75 eV, which is slightly higher than the value which gives the maximum output power. While the gap can be reduced by doping with Ge to give the optimum power, a marked enhancement of the efficiency cannot necessarily be expected.

In order to achieve a significant improvement in the efficiency, it is necessary to take a cell structure which allows the use of light energy in the wavelength range for which the available energy is lower. Attempts have been made to change the optical gap by alloying amorphous Si with C or Ge, and to construct stacked tandem cells with a cell of higher spectral sensitivity to shorter wavelengths placed on the incident-light side, followed by cells which are sensitive to increasingly longer wavelengths. The output powers from the individual cells can then be added together [4]. Fig. 6.6 shows an example of a three-stacked tandem cell, in which the relative contents of C and Ge are carefully changed, thus taking advantage of the degree of freedom which is allowed in the design of amorphous materials. This cell provides ⩾11% conversion efficiency [5], and is characterized by a reduced power loss at the lead wires owing to a higher voltage and lower current density than those of the single-layer cell.

Solar cells, particularly those used for power generation, which are exposed to strong illumination for a long period of time, require a high weather resistance. In the case of amorphous Si solar cells, a technological task of the utmost importance is to understand the mechanism of the reduction in the conversion efficiency as a result of the Staebler–Wronski effect, and to prevent this reduction, as well increasing the initial efficiency.

Efforts are being made to improve the cell stability, not only by making the film itself immune to optical aging, but also by specific design of the solar cell structure. The photo-induced drop in the cell efficiency may be attributable to the curtailment of the carrier lifetimes. If the thickness of the i-layer film is

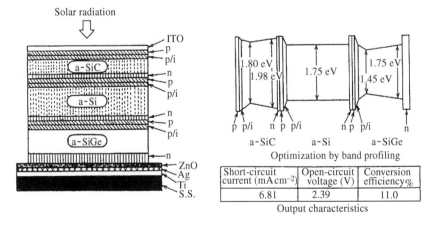

Fig. 6.6 A three-stacked tandem solar cell composed of a-SiC/a-Si/a-SiGe (reproduced by permission of the Sharp Corporation)

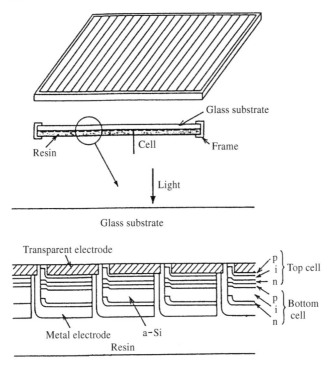

Fig. 6.7 Construction of a large-area solar cell (reproduced by permission of the Fuji Electric Company Ltd)

reduced, the shortening of the carrier life time does not cause the efficiency to deteriorate immediately, thus improving the cell stability. For this reason, it has been shown that the stacked-cell structure, in which the i-layer film of each cell can be reduced, has its stability as well as its efficiency improved.

Figure 6.7 shows an example of an a-Si/a-Si two-stacked tandem cell which achieves a 10% conversion efficiency with a light-receiving area as large as 30×40 cm^2 [6]. The degradation of this cell has been shown to be $\geqslant 10\%$. Moreover, the merits of amorphous materials has led to the steady development of solar cells for power generation, as exemplified by the trial manufacture of vast-area panels, as large as 40×120 cm^2, unrealizable for crystalline materials; the outer view for one of these is shown in Fig. 6.8.

The construction of stacked tandem cells, with the optical gap controlled by adding C or Ge, seems to be one of the most promising methods for cell fabrication. It should be noted that while the alloying of C or Ge allows ready control of the optical gap, reduction of the photoconductivity tends to degrade the film quality. In the example shown in Fig. 6.6, the device characteristics have been improved by changing the relative content of C or Ge. In order to achieve an efficiency better than 15% in the future, it will be necessary to further improve the film quality of amorphous SiC or SiGe.

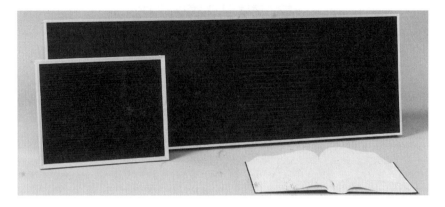

Fig. 6.8 An outer view of a large-area solar cell, fabricated by taking advantage of the technology now available for constructing large-area films of amorphous silicon: area of large cell, 40×20 (cm^2); area of small cell, 30×40 (cm^2) (reproduced by permission of the Fuji Electric Company Ltd)

Fig. 6.9 Amorphous silicon solar cells mounted on the sunroof of an automobile, fabricated by making use of the specific merits of amorphous materials, such as flexible workability, adaption to curved surfaces, and translucency, in the industrial design (reproduced by permission of the Mazda Motor Corporation)

Figure 6.9 shows a view of solar cells mounted on the sunroof of an automobile; these are formed on a curved surface, again taking advantage of the specific advantages of amorphous materials in the industrial design process.

On the other hand, the conversion efficiency of crystalline-silicon-based solar cells has reached 20% (and even higher), and therefore further enhancement of the efficiency in amorphous-silicon-solar cells is urgently required. While the theoretical upper limit of the conversion efficiency in amorphous-silicon solar cells is not yet completely established, owing to lack of knowledge of the ultimate properties of amorphous materials, it is by no means only a dream to be able to realize an efficiency higher than 20%.

6.2.2 SOLID-STATE PHOTOSENSORS

A variety of photosensors have been developed and commercialized through the utilization of photoconductive and photovoltaic effects, and are used for detecting various optical parameters such as light intensities, one-dimensional (1-D) line patterns, and two-dimensional (2-D) image patterns. If combined with color filters, the photosensor can be used as a color sensor.

In the case of photosensors equipped with a scanning mechanism, there are two modes of signal reading, i.e. 'cumulative', where the electrical signal output of each component is determined by an integral of the light intensity absorbed during the time interval between successive read-outs, and 'non-cumulative', in which the electrical signal output is proportional to the light intensity at each of the signal read-outs.

The cumulative type is useful in such applications as the 2-D image sensor, which contains a great number of components and where the read-out time is shorter than the read-out interval, because light incident during the read-out interval can be used. In the carrier-implant-preventing structure often used in the cumulative type, the quantum efficiency is equal to 1, i.e. the current multiplying factor is 1, under ideal conditions where as many carriers as absorbed photons are completely collected by the electrode.

On the other hand, in the injection-current-control structure which is frequently used in the non-cumulative type, carriers can be multiplied effectively, thus achieving a current-multiplying factor as high as 10^3. In this case, however, the recombination and extinction of injected carriers takes longer, the response is slower, and the linearity of the electrical output with respect to the light intensity is rather poor. This type is suitable, therefore, for those applications where the number of components is small or only one, linearity is not strictly required, and a slower scanning rate is acceptable.

(a) Color sensors

The structure of a color sensor is basically the same as that of the solar cells described previously in Section 6.2.1, with the load resistance reduced to that of the constant-current region, and used mainly in the short-circuited state. As the spectral sensitivity of a-Si:H is similar to that of the human eye, a solar cell can be converted into a color sensor by combination with three color filters. As is evident from the properties of solar cells, a good linearity is ensured between the sensor short-circuit current and the incident light intensity, thus giving a dynamic range as wide as six orders of magnitude.

(b) Contact-type linear image sensors

At the present time, the use of the facsimile machine is wide spread is, not only in offices, but also in the home. For this application, it is necessary to reduce the size of the image sensor in order to read the original document. While the contact-type linear image sensor itself is larger than the (charge-coupled device)

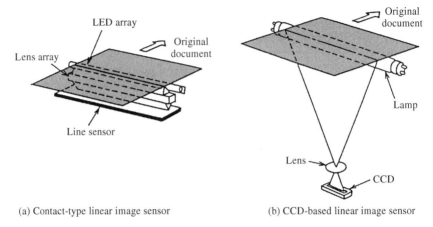

(a) Contact-type linear image sensor   (b) CCD-based linear image sensor

Fig. 6.10 A contact-type linear image sensor arrangement (a) for use in facsimile machines; another image sensor set-up which includes a CCD and a focusing lens system (b) is shown for comparison purposes

(CCD), the system can be reduced in size since a bulky focusing system is not needed in the former.

Figure 6.10 shows a schematic representation of contact-type linear image scanner [7]. With a light-emitting diode (LED) array as the light source, the intensity of the light reflected from an original document is transmitted by a lens array to a photodiode array, based on a-Si:H, which acts as the detector. These photodiodes are aligned at a density of approximately 10 units per mm. An array as long as 300 mm is used to handle A-3 sized sheets of paper. Usually, the cumulative mode is adopted for ensuring an adequate scanning rate and sensitivity. The electrical output is obtained by scanning with an external crystalline Si-based large-scale integrated (LSI) device.

(c) Line sensors for X-ray computerized tomography

Owing to rapid progress in its technology, computerized tomography (CT) based on the use of X-rays (see Fig. 6.11) is being increasingly utilized in the medical field. The X-ray line sensor for this application has evolved from the initial gas-ionization-chamber array design to one which now uses a combination of scintillators with crystalline Si photodiodes, with miniaturization being one of the aims of thus solid-state device. However, the size of the crystalline Si photodiode array is limited by the dimensions of the necessary Si-wafers, which are a few tens of centimeters in size at the largest. Moreover, owing to the limitations of the scintillator geometry and the need to arrange the sensors in a circular arc, the sensor system is currently constructed by laying out sets of ten sensor arrays a few centimeters in length, and with a few tens of channels for each, on a polygonal mount. However, this design is rather impractical from the cost point of view, because crystalline Si photodiode arrays are expensive,

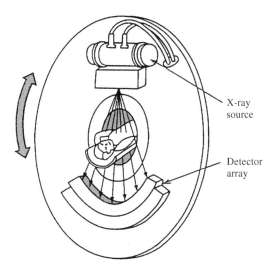

Fig. 6.11 Schematic representation of a detector system used in X-ray computerized tomography

and it is technically difficult to construct a sensor array of uniform sensitivity by combining the scintillators with the crystalline Si photodiodes in precise alignment.

Attempts have been made, therefore, to replace the crystalline Si photodiode array with an a-Si:H photodiode array. Initially, a (p-i-n)-type a-Si:H photo-diode array, similar to that used for solar cells, was attached to a thin glass substrate, and then combined with various scintillators in order to assess its characteristics.

Figure 6.12 shows a typical design for this type of detector, in which the scintillator is integrated with an a-Si:H photodiode array, by building the latter on to the scintillator, which is used as the substrate. Owing to a good match with the emission spectrum of the scintillator, the sensitivity and the signal-to-noise (S/N) ratio of a-Si:H photodiode array are almost as good as those of the crystalline-Si-sensor array [9]. Precision alignment of the scintillators with the photodiode arrays is not needed, and the sensitivity is adequately uniform in nature. It has also been demonstrated that potential aging caused by continuous irradiation with X-rays poses no problem. This type of system is expected to become increasingly used in the future on account of its comparatively low cost.

(d) Two-storied, two-dimensional image sensors

A charge-coupled device (CCD)-type 2-D image sensor based on crystalline LSI technology is often used as an image pick-up device, where the latter constitutes the very heart of the home video camera. In order to improve the resolution of the camera, it is necessary to increase the number of pixels, while

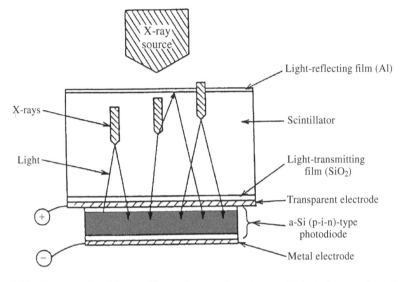

Fig. 6.12 Integrated solid-state X-ray detector based on a-Si (p-i-n)-type photodiodes [9, reproduced with permission. © 1990 IEEE]

Fig. 6.13 Two-storied, two-dimensional which uses image sensor a-Si and a CCD (reproduced by permission of the Toshiba Corporation)

for improving the sensitivity, the photosensitive area must be increased. However, if the number of pixels is increased, then the effective light-receiving area of the sensor is correspondingly reduced.

In order to improve the sensitivity of the CCD- or metal-oxide semiconductor (MOS)-type 2-D image sensor, a proposed design involves the building up of a 2-D image sensor of a two-storied construction in which the CCD- or MOS-type sensor is used as the scanning component while a-Se or a-Si:H mounted over the entire top area is used as the sensor component.

Figure 6.13 shows an example of a two-million-pixel 2-D image sensor which uses a-Si:H as the sensor element and the CCD as the scanning component. The specifications for the sensor element and the cross-sectional structure of the pixel unit are shown in Table 6.2 and Fig. 6.14, respectively. It has been demonstrated that an a-Si:H layer, without the isolation of each pixel, is characterized by a low dark current and high resolution (1000 lines or better), which is little affected by transverse migration of the photo-induced carriers, as well as posessing a high sensitivity, and is therefore, expected to find applications as a component in the new generation of high-definition television systems [10].

Table 6.2 Specifications for an a-Si/CCD two-storied, two-dimensional image sensor $(2 \times 10^6$ pixels)

Parameter	Value[a]
Chip size	16.2 (H) × 10.5 (V) mm
Sensor area	14.0 (H) × 7.8 (V) mm (compatible to 1" optical system)
Aspect ratio	16:9
Effective number of pixels	1920 (H) × 1036 (V)
Pixel size	7.3 (H) × 7.6 (V) μm
Saturation current	2000 nA (2×10^5 electrons per pixel)
Sensitivity	210 nA lx^{-1}
Dynamic range	72 dB
Horizontal liminal resolution	1000 lines
Smear	−100 dB

[a]H, horizontal; V, vertical

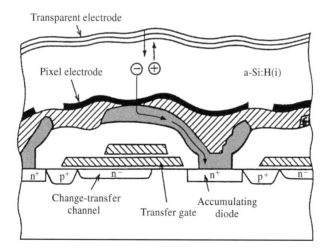

Fig. 6.14 Cross-section through a cell unit of an a-Si/CCD two-storied, two-dimensional image sensor

6.2.3 FIELD-EFFECT (THIN-FILM) TRANSISTORS FOR LIQUID CRYSTAL DISPLAYS

While the applications described in Sections 6.2.1 and 6.2.2 are based on photo-electric effects, those to be dealt with in this section are characterized by the utilization of field effects which are not related to photonics.

Although a variety of image display elements have been proposed, studied and trial-manufactured, none of these have provided any real competition to the conventional TV cathode-ray tube (CRT) with respect to image quality, functionality and price. However, the CRT requires a depth which is as large as the image width, and therefore the development of a thin flat image-display device, as represented by a wall-hung television set, is urgently required. Recently, a liquid crystal display device, with an active matrix design based on a-Si:H thin-film transistors (TFTs) of the field-effect type as the switching component, has been shown to be of practical use [11]. This display is finding applications in small television receivers, projection-type displays, and computer terminal monitors. As distinct from the CRT, the display has a thickness which is less than 5 mm (including a glass substrate), and even when components such as driving circuits and fluorescent tubes for illumination are incorporated, it is still only 30 mm or so thick. As this thickness is expected to be reduced still further, it can be claimed that a positive step has been made towards the implementation of wall-hung television sets, flat-type personal computers and image display walls [12].

Fig. 6.15 A view of a typical display on a 9.4' TFT-driven liquid crystal color display unit

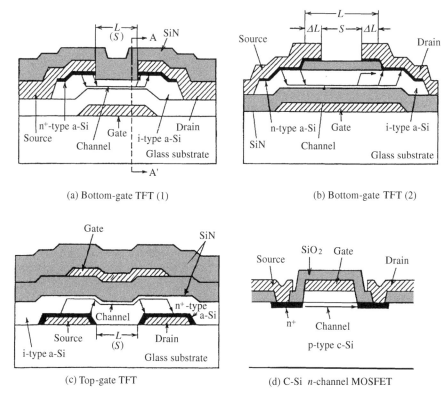

(a) Bottom-gate TFT (1)

(b) Bottom-gate TFT (2)

(c) Top-gate TFT

(d) C-Si n-channel MOSFET

Fig. 6.16 Typical examples of cross-sections through various types of a-Si TFTs (a–c); for comparison purposes, a cross-section of a c-Si MOS transistor is also shown (d) [13, reproduced with permission]

Figure 6.15 shows an example of a TFT-driven liquid crystal color display unit (480 (H) × 640 (W) × 3(colors) = ca 9.2×10^5 pixels).

Cross-sections through various types of a-Si TFTs used to drive each of these pixels are shown in Fig. 6.16(a–c) [13]. For comparison, a cross-section through an n-channel MOS transistor made from crystalline Si is shown in Fig. 6.16(d). While a channel for switching the electron current on or off is located in the same plane as the source/drain units in the c-Si MOS transistor, this channel is placed in an opposite plane to that of the source/drain units in the TFT models, because of limitations placed by the fabrication process. The design in which a gate electrode is positioned on the substrate side (Fig. 6.16(a,b)) is known as a 'bottom gate' type, and that in which the source/drain units are placed on the substrate side (Fig. 6.16(c)) as a 'top gate' type. In design (a), the number of working steps is reduced because i-type a-Si and n^+-type a-Si are formed successively. However, when the n^+-type a-Si is removed by etching, it is inevitable that some of the i-type a-Si will be lost. For this reason, it is necessary to make the i-type a-Si layer as thick as 1000 Å or more. On the other hand, design (b) needs a greater number of steps as the SiO_2 layer will be formed between layers of i-type and n^+-type a-Si; however, it does have the advantage of containing an i-type a-Si film as thin as 100 Å, and TFT characteristics which are only slightly affected by the leakage of light.

Figure 6.17(a) shows the gate voltage (V_G)–drain current (I_D) characteristics [14] of the a-Si TFT shown in Fig. 6.16(a), while Fig. 6.17(b) shows an energy diagram of the same a-Si TFT through the A–A$'$ section shown in Fig. 6.16(a) at gate voltage 0, where the broken lines represent the case of a positive bias. The i-type a-Si layer shown in Fig. 6.17(b) is of a weak n-type, and when a negative bias of ca −5 V is applied to the gate ($V_G \ll V_t$), the drain current I_D changes to an OFF current of the order of 10^{-14} A, as determined by specific resistance of the i-type a-Si; V_t is the threshold voltage.

When a positive bias is applied, the energy diagram of the TFT becomes as shown by the broken lines in Fig. 6.17(b), in which electrons are accumulated at the channel to give an exponential increase in the current. If the positive bias is raised to $V_D (\gg V_G - V_t)$, the drain current becomes $I_D = (W/2L)\mu C_{INS}(V_G - V_t)^2$, where W is the channel width, L the channel length, μ the mobility of an electron within the channel (field-effect mobility), and C_{INS} the gate-insulating-film capacitance per unit area. It is possible to obtain the values of μ and V_t from the measured value of $V_G - I_D$ by using this relationship. In the example shown in Fig. 6.17(a), $\mu = 1.1$ cm^2 V^{-1}s^{-1} and $V_t = 1.5$ V. The ON/OFF ratio of I_D becomes $\geqslant 10^8$ at $V_G \pm 10$ V.

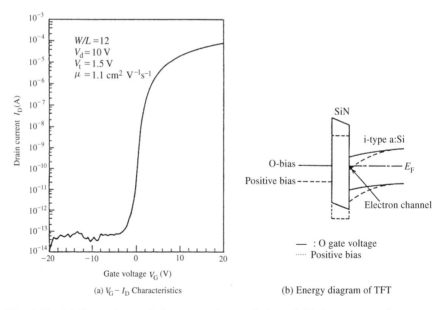

(a) $V_G - I_D$ Characteristics

(b) Energy diagram of TFT

Fig. 6.17 (a) Gate voltage–drain current characteristics and (b) the corresponding energy diagram of an a-Si TFT [14, reproduced with permission]

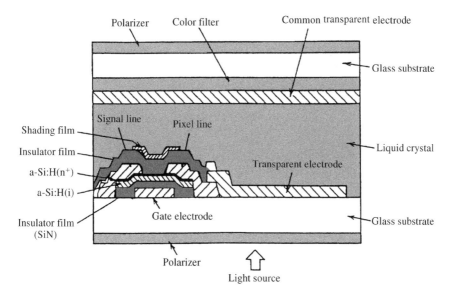

Fig. 6.18   Cross-section through a TFT–LCD cell unit

The a-Si TFT can be combined with an optical switch which is based on liquid crystals to form the cell unit shown in Fig. 6.18 [13]. The operation of this unit may be briefly described as follows. As shown on the right side of Fig. 6.18, the cell unit is composed of the following elements, reading from the bottom: white light source, polarizer, glass substrate, transparent electrode, liquid crystal, common transparent electrode, color filter, glass substrate and polarizer. The display is to be viewed from the top. The bottom side transparent electrode is connected to the drain electrode of the TFT, while the common transparent electrode is grounded.

As described above, the drain voltage is varied depending upon the image signal in the TFT. When the drain voltage is equal to 0, the drain electrode is at the same potential as the common transparent electrode, and the liquid crystal molecules are oriented, i.e. twisted by 90°, as shown in Fig. 6.19(a). For this reason, when a light beam passes through the liquid crystal, its direction of polarization is rotated by 90°, which is orthogonal to the polarization direction of the top polarizer, and therefore no light is transmitted. When a drain voltage is applied, the liquid crystal molecules are oriented perpendicularly (as shown in Fig. 6.19(b)), with the result that the polarization of the light beam is unaffected and thus passes through the top polarizer. A color display is obtained by the deposition of a pixel set equipped with three color filters.

Figure 6.20 shows the circuit diagram of a display matrix for a liquid crystal color display unit [14]. In this figure, R (red), G (green) and B (blue) represent a three-color pixel set. A horizontal row of TFT gates and a vertical column of TFT sources are connected to the drive circuit through their respective gate lines and

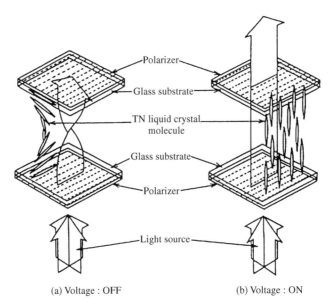

(a) Voltage : OFF   (b) Voltage : ON

Fig. 6.19   Operation of the liquid crystal (TN-type) display component

data lines. When a voltage is applied to gate line No. 1, the N TFTs (from 11 to
$1N$) which are connected to this gate line are turned ON, and the corresponding
signal currents are fed simultaneously from the data lines into the gate lines
Nos $1-N$ to drive the liquid crystals. This operation is repeated sequentially up
to the Mth gate line, thus constructing a display image (sequential line scan
method). A motion picture can be builtup by repeating these steps. Since each
pixel is required to keep its current state up to the next drive, the signal charge
is maintained by the use of the electrostatic capacitances of the liquid crystals
and additional storage capacitors.

As greater drive powers and faster operation rates are needed, a number of
c-Si-based LSI units are installed around the substrate for the driving circuits of
the gate and data lines. For these applications, development of TFTs with even
higher drive powers and faster operation rates is urgently required. If these TFTs
can be formed on the same substrate, the cost may be reduced dramatically.
Attempts have been made to markedly improve the mobility, by reducing the
hydrogen content in a-Si and by utilizing microcrystalline or polycrystalline films.

Microcrystalline or polycrystalline films have carrier mobilities which are 10
to 100 times higher than those of amorphous films. For this reason, attempts
have been made to use such films, not only for the drive circuits (in place of
the c-Si LSI units, but also for all of the major components, including the pixel
switches. However, most of the commercially available products still utilize low-
cost amorphous film devices, on account of the need for expensive substrates to
cope with the higher processing temperatures of the 'crystalline' films, and the
higher OFF-current levels of the TFTs which use them.

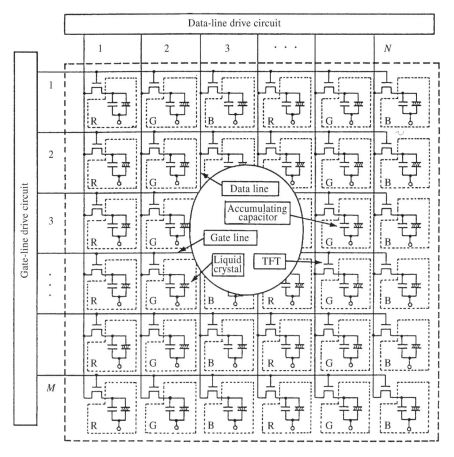

Fig. 6.20 The circuit diagram of a display matrix for a liquid crystal color display unit [14, reproduced with permission]

When the i-layer of an a-Si TFT absorbs light, photocarriers are produced which change the TFT characteristics, and therefore a shading film needs to be provided (as shown on the left side of Fig. 6.18. The photosensitivity of the TFT may be effectively suppressed by using an i-layer of a-Si as thin as 100 Å, as previously described with reference to Fig. 6.16(b).

It is equally important to secure long-term stability, as well as suppressing the photosensitivity. In particular, it is essential to suppress the shift of V_t (as shown in Fig. 6.17(a)). Although not discussed further here, the major causes of the V_t shift are thought to be charge trapping by the interface levels of the a-Si/SiN or by the levels generated by breaking of the weak Si−Si bonds in a-Si in the neighbourhood of the interface. However, as the physical mechanism underlying this effect is not yet fully understood, further studies are required. It is possible that a factor common to the Staebler–Wronski effect, described above with reference to solar cells (Section 6.2.1), may be involved.

While a few problems still remain to be solved, the commercialization of TFT-based liquid crystal display devices is making rapid progress, and currently represents the largest area of a-Si application. However, the cost of TFT-based liquid crystal display devices is an order of magnitude higher than that of CRT-based displays of the same size, and therefore the need to develop new technology for reducing this cost is of prime importance.

6.2.4 PHOTORECEPTOR DRUMS FOR ELECTROPHOTOGRAPHY AND LASER-BEAM PRINTERS

The use of copying machines has recently shown a remarkable growth. The basic operation of the copying machine is based on the Carlson process, in which carbon particles are thermally fixated on to a plain paper sheet (Fig. 6.21). In this process, Se, a-Si or organic photoconductors are used as the photoconductive materials for forming charge patterns by either photo-image patternerning or laser-beam scanning.

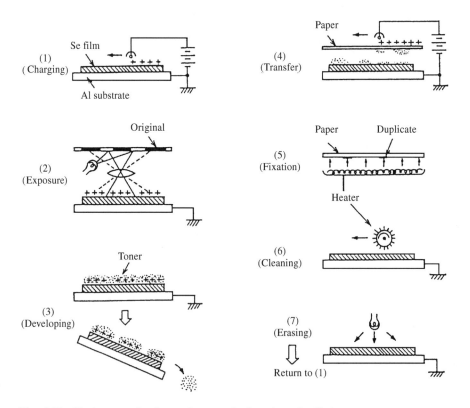

Fig. 6.21 Basic stages in electrophotography based on the Carlson process

In the actual copying machine drum, a photoreceptor layer, tens of microns thick, is deposited on to a cylindrical aluminum substrate through vapor deposition in vacuum (for Se), or through glow discharge decomposition of monosilane (SiH_4) (for a-Si).

The surface of the photoreceptor layer is charged by a corona discharge or similar means up to a surface potential of *ca* 1000 V, as shown in Fig. 6.21(1). Then, an original drawing is projected on to the drum surface, through a lens system in the case of electrophotography (2), or is written via laser-beam scanning in the case of a laser-beam printer. While the surface potential remains constant in the dark areas, it is reduced in the bright areas as the surface charges are neutralized by the photogenerated carriers which correspond to the brightness.

Consequently, the light image is converted into a charge image. When a thermo-adhesive powder of an insulator containing carbon particles, called the toner, is spread over the charge image, this powder is adsorbed electrostatically in those areas where the surface charges remain (3). If a sheet of paper is then placed close to the drum and an electric field is applied from the back of the paper, the charged image of the toner is transferred to the paper (4). As the paper is heated, the toner is thermally fixed on to the paper to complete the copying process (5). Finally, the drum surface is cleaned (6) and then illuminated over the entire surface in order to eliminate any residual charge (7), thus returning to its original state (1).

In order to repeat a cycle of these steps over 0.1 to 10 s, the photoreceptor drum must possess the following properties:

(1)   Charge potential: a surface potential of around 10^3 V provided by the corona discharge must be held in the dark for tens of seconds (or even longer) without any decay, and the residual potential after resetting the illumination must be suitably low.

(2)   Photosensitivity: the quantum efficiency must be adequately high, with the spectral sensitivity being compatible with the wavelength of the recording light.

(3)   Resolution: the resistance needs to be at such a level that transverse migration of the charge, which can cause image blurring, is effectively suppressed.

(4)   Abrasion resistance: the photoreceptor drum, which is constantly in direct contact with toner, paper, cleaning brushes, etc., must have a high abrasion resistance.

(5)   Weather resistance: as the surface of the drum is exposed to the atmosphere, it must display a high resistance to humidity and contamination.

In order to meet these requirements, composite multilayer films are used for the photoreceptor layers [15]. Fig. 6.22 shows an example of a negatively charged photoreceptor layer which is based on a-Si materials, consisting of a surface-protecting layer, a carrier-generating layer, a carrier-drifting layer and a blocking layer. The surface-protecting layer is highly resistant to abrasion and

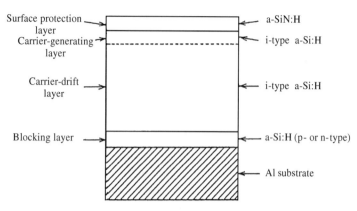

Fig. 6.22 Cross-section through a photoreceptor drum used for electrophotography which is based on a-Si:H [15, reproduced with permission]

weathering, and has excellent charge-holding properties, thus prevent degradation of the resolution due to transverse migration of charge, while allowing light to pass through to the carrier-generating layer without being absorbed. In this example, a-SiN:H is used. A film of a-Si:H, doped with a amount of boron, i.e. electronically intrinsic a-Si:H, is used as the carrier-generating layer. This layer absorbs the incident light, and efficiently generates electron–hole pairs, for which charges of the opposite sign to the applied charge (holes in this case) act to neutralize the negative charges, while those of the same sign (electrons in this example) flow towards the Al substrate via the carrier-drifting and blocking layers. For this reason, the carrier-drifting layer is required to have a high mobility for carriers of the same sign as the applied charge (i.e. electrons), and to contain no traps. Undoped a-Si:H, of the weak n-type, with a high mobility for electrons and containing few traps, is used for this layer. The blocking layer is n-type a-Si:H, which stops (blocks) the carriers, in particular those of an opposite sign to the applied charge (holes in this example) from being ejected from the Al substrate. Without this blocking process, carriers (holes in this case) implanted from the Al substrate, would pass through the carrier-drifting layer and neutralize the applied charge, thus leading to deterioration of the dark-potential holding properties.

Figure 6.23 shows changes in the surface potential under illumination with 600 nm light of varied intensity, while Fig. 6.24 shows the photoconductivity gain, i.e. the wavelength dependence of the quantum efficiency [15]. This efficiency is nearly 100% at wavelengths shorter than 700 nm. The light absorbance shown by the broken line indicates that light with a wavelength of about 700 nm (or longer) passes through the layer to reduce the photoconductivity gain. Over such a long wavelength range, the light generates carriers, not only in the carrier-generating layer, but also uniformly in the carrier-drifting layer. Both electrons and holes can act as drifting carriers

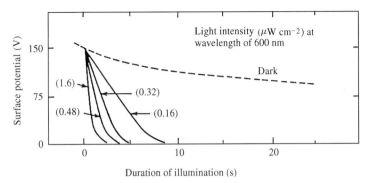

Fig. 6.23 Changes in surface potential as a function of illumination intensity [15, reproduced with permission]

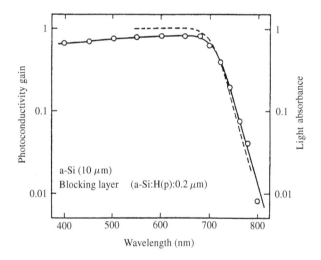

Fig. 6.24 Dependence of photoconductivity gain on wavelength; for comparison purposes, the broken line shows the corresponding optical absorption [15, reproduced with permission]

Figure 6.25 shows an example of an a-Si photoreceptor drum which is used in a laser-beam printer.

6.2.5 AMORPHOUS SELENIUM VIDICON-TYPE IMAGE PICK-UP TUBES

As described earlier in Chapter 1, amorphous selenium (a-Se) is a material containing chain- or ring-like structures based on atomic bonds of coordination 2, and is characterized by flexibility and lower localized-level densities. Fig. 6.26 shows an example of a vidicon-type image pick-up tube using a-Se film, which has a cross-sectional construction of the avalanche multiplying type

Fig. 6.25 An example of an a-Si photoreceptor drum used in the laser-beam printer (reproduced by permission of Hitachi Ltd)

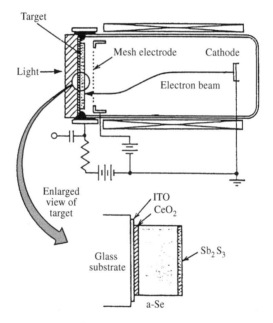

Fig. 6.26 Construction of a vidicon-type image pick-up tube which uses a-Se [16, reproduced with permission]

(to be described below) [16]. The device consists of an evacuated glass tube containing a photoconductive film, called a target, and a cathode for emitting electrons. Although the details are not shown in the figure, it is also provided with an electrostatic or electromagnetic deflection mechanism for scanning the electron beam two-dimensionally, both inside and outside the tube.

The target consists of a glass substrate stacked with $(In-Sn-O)$ (ITO) as a transparent electrode, CeO_2 as a blocking layer, a-Se as a photoconductive layer, and porous Sb_2S_3 for suppressing any secondary electron emission. The ITO serves as a signal output terminal, and has a positive voltage of some tens of volts, applied via a load resistance. On the other hand, the cathode shown on the right-handside of the figure is held at ground potential; the Sb_2S_3 surface is also held at ground potential (similar to the cathode), in the dark state, by scanning with an electron beam which is emitted from the cathode. The photoconductive film of a-Se is a few microns thick, with an applied electric field of ca 10^5 V cm^{-1}.

It is essential that the carriers, i.e. holes injected from the ITO and electrons injected from Sb_2S_3, do not augment the dark current, and ITO/CeO_2/a-Se and a-Se/Sb_2S_3 junctions are therefore used to prevent this carrier injection. In order to operate the camera in the visible light region, as in the case of the color camera, a ternary material, $Se-Te-As$ (to be designated simply as an a-Se layer, for the sake of convenience) is used, with Te added to compensate for the inadequate sensitivity to red light, plus As for enhancing the thermal stability and suppressing crystallization. So as to avoid other properties from being adversely effected by these additives, both the content and the site of addition are controlled to atomic layer precision.

When light from the left side is accepted to form a light image on the target, electron–hole pairs are generated in proportion to the light intensity at the a-Se layer which is closest to the incident light, and neutralize the charges in those areas of the target surface which correspond to the light image. While this neutralization process continues as long as the light is received (integrating effect), when the scanning electron beam hits a particular pixel for a second time, the photoconductive film surface is charged up to the ground potential. On this occasion, the current in the beam circuit is proportional to the accumulated charge, and is displayed as signal output. Scanning with the electron beam is continued sequentially in order to convert the two-dimensional image information into time-dependent one-dimensional electrical signals. As the incident light is integrated, a photoelectric conversion of high sensitivity and low noise is realized.

Figure 6.27 shows the current–voltage characteristics of the a-Se vidicon-type image pick-up tube, indicating the excellent blocking properties of the ITO/CeO_2/a-Se, a-Se/Sb_2S_3 junctions, for applied voltage up to 200 V. The saturation area of the photocurrent corresponds to a quantum efficiency 1, where all of the generated electron-hole pairs contribute to the signal current. The image pick-up tube which makes use of this saturation-area feature has practical applications in the cameras used for television broadcasting.

Owing to the increasing popularity of home video cameras, the performance of the two-dimensional solid-state image pick-up device of the CCD type, based on the LSI technology of crystalline Si, is improving day-by-day. On the other hand, in the face of practical applications of high-definition television broadcasting technology, the demands for both high resolution and high sensitivity continue to be emphasized. However, in view today's technological levels, it

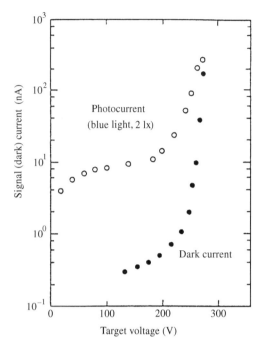

Fig. 6.27 Current–voltage characteristics of the avalanche multiplying image pick-up tube [16, reproduced with permission]

seems that improvements in light utilization have reached their limits, and any further significant enhancements in sensitivity seem unlikely.

However, the discovery of carrier multiplication in a-Se films has provided a breakthrough for the development of an ultrasensitive image pick-up tube. This carrier multiplication has been found to result from the avalanche multiplication effect in amorphous materials, which has been studied very little up until recently [17]. Moreover, this multiplication process has been shown to occur uniformly in films, and the recording of images in the accumulation mode ensures a very low noise level.

The performance of an avalanche-multiplication-type image pick-up tube using a-Se is shown in Fig. 6.28. In contrast to a conventional image pick-up tube which uses the saturation area, the positive voltage that is applied through the load resistance can be as high as hundreds of volts.

In the current–voltage characteristics shown in Fig. 6.27, if the applied voltage is higher than 200 V, the photocurrent increases rapidly, while the dark current remains unchanged. As described above, the saturation area for an applied voltage of 200 V or lower corresponds to a quantum efficiency of 1, and it is evident that for the area on the right-handside of Fig. 6.27 where the photocurrent is enhanced, the quantum efficiency should be exceptionally high [16]. If the a-Se film is ⩾20 μm, and the applied voltage is precisely controlled without any fluctuations,

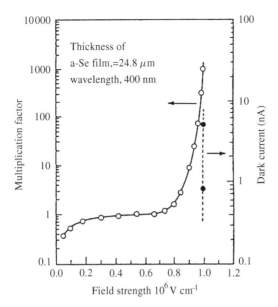

Fig. 6.28 Performance of an ultra sensitive image pick-up tube of the avalanche multiplication type which uses a-Se

then a uniform and stable ultrasensitivity (as high as 10^3) is obtained, as shown in Fig. 6.28 [16]. In this way, a low-noise image is achieved without degrading the resolution nor increasing the 'after-image'.

Let us now consider the mechanism for this extra-high sensitivity and very low noise level. In order to investigate the avalanche multiplication process in a-Se films, a cell with a simple sandwich structure was constructed with Al/CeO_2 being used as the hole-blocking electrode, and Au as the electron-blocking electrode (see Fig. 6.29) [18]. While applying a voltage to the Al/CeO_2 (anode) and Au (cathode), changes in the current were measured, using light of wavelength 400 nm, projected from either the side of the Al/CeO_2 or that of the Au. For the former, the light is absorbed by a very thin layer on the incident side of the a-Se, thus allowing only holes to drift, while in the latter case, only electrons are allowed to drift.

Figure 6.30 shows the dependence on field strength of the current density under conditions of hole-drift in the cell. As in the case of the image pick-up tube, the saturation area corresponds to a quantum efficiency of 1, and a steep rise in the current is seen at electric fields above a certain value. As the film thickness of the a-Se is increased, a corresponding increase in the steepness of the current versus field strength curve is observed. This trend is in contrast to the changes expected as a result of the space-charge limited current or photo-induced charge injection from the electrode, and is most likely due to increases resulting from the avalanche-multiplication effect.

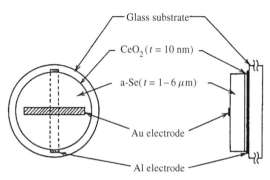

Fig. 6.29 A schematic representation of the cell used in the sandwich-structure-type a-Se avalanche photodiode [18, reproduced with permission]

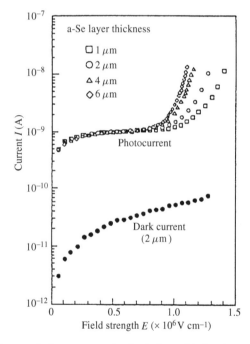

Fig. 6.30 Dependence of photocurrent (under hole-multiplication conditions) and dark current on field strength and thickness of the a-Se layer [18, reproduced with permission]

If this rise is considered to be caused by avalanche multiplication, and the ionization factor due to holes or electrons is then estimated, the curves shown in Fig. 6.31 are obtained. The figure includes the ionization factors of some well-known crystalline materials [19], and therefore the particular features of a-Se can be clearly identified. Particular attention should be paid to the fact that the ionization factors for the holes are much greater than those of the electrons,

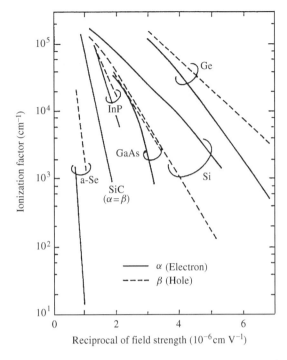

Fig. 6.31 Ionization factors of a-Se and various crystalline semiconductors [19, reproduced with permission]

and that the ratio of the ionization factors (holes/electrons) can be as large as 100. This feature is particularly favourable for image pick-up tubes in which multiplication by using holes is the operational process.

Since the ratios of the ionization factors (holes/electrons) are large, the cycled multiplication due to the reverse drift of the electrons generated by the multiplication process is suppressed. This prevents the development of an after-image and excessive increases in noise due to fluctuations in the multiplication, thus resulting in an image with a high signal-to-noise (S/N) ratio.

Figure 6.32 shows an image taken by an a-Se avalanche-multiplication-type image pick-up tube, compared with an image taken with a conventional device, using the same conditions (200 lx and lens stop F4) [16]. The a-Se avalanche-multiplication-type image pick-up tube with a multiplication factor (see of $\geqslant 10^3$ Fig. 6.28), is so sensitive that it can recognize characters at illuminations which are as low as 0.1 lx, thus far surpassing the visual capabilities of the human eye.

The ionization process caused by avalanche multiplication is stochastic in nature, involving noises caused by fluctuations in the multiplication factor. The S/N ratio which is actually measured with this camera tube is a few times higher than that expected in the ideal state where noise due to fluctuations is assumed to be absent. However, the spectral intensity of the noises generated

Fig. 6.32 Images taken by using an a-Se image pick-up tube under avalanche-multiplication conditions (left) and under normal conditions (right); in both cases, an illumination of 200 lx and a camera lens stop F4 were used [16, reproduced with permission]

in the cell structure of Fig. 6.29 is the same as that observed for crystalline materials.

One of the causes of a high S/N ratio can be identified in the multiplication characteristics shown in Fig. 6.28, where the multiplication factors rise so steeply with respect to the field strengths that any minor increases in the multiplication factor at a certain time point and a certain site make the surface potential lower than the mean, and also reduce the internal field below its mean value. For this reason, the multiplication factor of the incoming carrier is slightly reduced. This means that operation in the accumulation mode must involve a negative feedback in order to reduce the noise, and thus enhance the S/N ratio, so reducing the noise level to a value which is much lower than expected under an ideal state without any negative feedback.

The avalanche-multiplication-type image pick-up tube has a number of commercial applications, and is currently being used for astronomical observations in the night-time sky and for television broadcasts which involve fast-moving sports scenes.

In order to expand the wavelength range to be covered by a image pick-up tube, attempts have been made to replace the glass substrate and incident window with beryllium, and the resulting image pick-up tube, which is sensitive to X-rays, has been trial-manufactured.

It is of great interest to study the avalanche-multiplication effect in other amorphous materials, such as a-Si:H. While a multiplication of the photocurrent has been observed in both a-SiC:H and a-Si:H, quantitative confirmation of avalanche multiplication has so far only been achieved with a-Se.

References

1. D. E. Carlson and C. R. Wronski, *Appl. Phys. Lett.*, **28** (1976) 671; D. E. Carlson, *IEEE Trans. Electron. Devices*, **ED-24** (1977) 449.

2. H. Okamoto, Y. Nitta, T. Adachi and Y. Hamakawa, *Surf. Sci.*, **86** (1979) 486.

3. Y. Kuwano and S. Nakano, *Amorphous Semiconductor Technologies and Devices*, JARECT 16 Ohmsha and North-Holland, Amsterdam (1984) 222.

4. Y. Hamakawa and H. Okamoto, *Amorphous Semiconductor Technologies and Devices*, JARECT 16 Ohmsha and North-Holland, Amsterdam (1984) 200.

5. Y. Nakata, H. Sannomiya, S. Moriuchi, Y. Inoue, K. Nomoto, A. Yokota, M. Itoh and T. Tsuji, *Optoelectronics—Devices and Technologies*, **5** (1990) 209.

6. T. Yoshida, A. Fujikake, H. Fujisawa, S. Saito, T. Sasaki, Y. Ichikawa and H. Sakai, *Proceedings of the 10th EC Photovoltaic Solar Energy Conference*, Lisbon (1991).

7. T. Yamada, M. Morita and T. Nagata, *Hitachi Rev.*, **38** (1989) 309.

8. H. Yamamoto, T. Baji, H. Matsumaru, Y. Tanaka, K. Seki, T. Tanaka, A. Sasano and T. Tsukada, *Extended Abstracts of the 15th Conference on Solid State Devices and Materials*, Tokyo (1983) 205.

9. T. Takahashi, H. Itoh, T. Shimada and H. Takeuchi, *IEEE Trans. Nucl. Sci*, **37** (1990) 1478.

10. M. Sasaki, R. Miyagawa and S. Manabe, *Extended Abstracts of the 22nd Conference on Solid State Devices and Materials*, Sendai (1990) 705.

11. M. J. Powell, *IEEE Trans. Electron. Devices*, **36** (1989) 2753.

12. M. Tsumura, M. Kitajima, K. Funahata, Y. Wakai, R. Saito, Y. Mikami, Y. Nagae and T. Tsukada, *SID '91 Digest*, (1991) 215.

13. T. Tsukada, *Flat Panel Displays*, Nikkei BP (1991) 88; T. Tsukada, *Liquid Crystal Displays and Related Materials*, lecture on optoelectronics presented at Kansai Chapter, Institute of Electronics, Information and Communication Engineers, Osaka (1990) (both in Japanese).

14. T. Tsukada, *Digest of the 49th Annual Device Research Conference Boulder*, (1991) IA-3.

15. T. Kawamura, N. Yamamoto and Y. Nakayama, *Amorphous Semiconductor Technologies and Devices*, edited by Y. Hamanaka, JARECT, Volume 6, Ohmsha and North-Holland, Amsterdam (198) 325.

16. K. Takasaki, K. Tsuji, T. Hirai, E. Maruyama, K. Tanioka, J. Yamazaki, K. Shidara and K. Taketoshi, *MRS Symp. Proc.*, **118** (1988, Spring) 387; K. Tsuji, T. Ohshima, T. Hirai, N. Gotoh, K. Tanioka and K. Shidara, *MRS Symp. Proc.*, **219** (1991, Spring) 507.

17. K. Tanioka, J. Yamazaki, K. Shidara, K. Taketoshi, T. Kawamura, S. Ishioka and K. Takasaki, *IEEE Electron. Devices Lett.*, **EDL-8** (1987) 392.

18. T. Ohshima, K. Tsuji, K. Sameshima, T. Hirai, K. Shidara and K. Taketoshi, *Jpn J. Appl. Phys.*, **30** (1991) L1071.

19. S. M. Sze, *Physics of Semiconductor Devices*, 2nd Edn, John Wiley & Sons, New York (1981), Ch. 10.

INDEX